C语言程序设计与应用

主　编　汪天富　董　磊

副主编　郭文波　曾凡均

電子工業出版社
Publishing House of Electronics Industry
北京·BEIJING

内 容 简 介

Visual Studio 是一个基本完整的开发工具集，包括了整个软件生命周期中所需要的大部分工具，如 UML 工具、代码管控工具、集成开发环境等。

本书基于 Visual Studio Community 2019 平台，通过 14 个实验介绍 C 语言程序设计与应用，其中 8 个实验为秒值-时间值转换实验，2 个实验基于多媒体定时器设计电子钟，其余 4 个实验与实际应用相关。

本书配有丰富的资料包，包括 C 语言例程、软件包、PPT 和视频等。资料包会持续更新，下载链接可通过微信公众号“卓越工程师培养系列”获取。

本书既可以作为高等院校相关课程的教材，也可以作为 C 语言开发及相关行业工程技术人员的入门培训书。

图书在版编目（CIP）数据

C 语言程序设计与应用 / 汪天富，董磊主编. —北京：电子工业出版社，2021.8
ISBN 978-7-121-41769-6

Ⅰ. ①C… Ⅱ. ①汪… ②董… Ⅲ. ①C 语言－程序设计 Ⅳ. ①TP312.8

中国版本图书馆 CIP 数据核字（2021）第 159923 号

责任编辑：张小乐
印　　刷：北京七彩京通数码快印有限公司
装　　订：北京七彩京通数码快印有限公司
出版发行：电子工业出版社
　　　　　北京市海淀区万寿路 173 信箱　　邮编：100036
开　　本：787×1092　1/16　印张：10.5　字数：269 千字
版　　次：2021 年 8 月第 1 版
印　　次：2024 年 8 月第 5 次印刷
定　　价：38.00 元

凡所购买电子工业出版社图书有缺损问题，请向购买书店调换。若书店售缺，请与本社发行部联系，联系及邮购电话：（010）88254888，88258888。

质量投诉请发邮件至 zlts@phei.com.cn，盗版侵权举报请发邮件至 dbqq@phei.com.cn。

本书咨询联系方式：（010）88254462，zhxl@phei.com.cn。

前　言

市面上的C语言教材非常多，大多数初学者学习C语言的方法是买或借一本C语言教材，先学习语法，甚至深究其中的语法，部分有实战意识的初学者会自行搭建开发环境，然后根据书中的例子编写一些程序。按照这种方法学完后通常会出现一个问题，似乎掌握了一些C语言知识，也可以编写一些小程序，但是无法使用C语言解决实际问题，例如，编写一些实用算法（如循环队列等），或编写单片机程序。

为什么会出现这样的问题？我们先分析一下唐僧团队取经历程，师徒四人西天取经，沿途收拾妖怪只是为了解决障碍，而不是要将天下的妖怪全都消灭。因此，孙悟空并没有沉迷于要把所有妖怪斩草除根，托熟人、找关系、搬救兵，将这些妖怪收走也可以。同理，在学习C语言时，要将其视为工具，目的是解决现实中的具体问题，在解决问题的过程中，遇到不懂的知识点便去有针对性地学习，而且不深陷于知识点中，问题解决后就应尽快返回到“取经之路”上。

本书是一本介绍C语言程序设计的书，严格意义上讲，也是一本实训手册。本书基于Visual Studio Community 2019集成开发环境，第1章介绍开发环境的安装和配置，磨刀不误砍柴工，磨好开发环境这把“刀”，才能高效地砍下第2～15章这些“柴”。其中，第2章介绍秒值-时间值转换的基础实验。第3～9章分别以数组、函数、枚举、指针、结构体、结构体指针和多文件的方式实现第2章的功能，这种通过不同语法完成相同实验的方式，有助于读者把精力聚焦在C语言的语法上。第10～11章引入了多媒体定时器的概念，不仅要求能够将秒值转换为时间值，还要让秒值递增计数，并通过printf函数每秒打印一次时间值。掌握了这些基本语法后，就可以尝试解决现实中的一些具体问题。第12章是一个算法设计的实例（循环队列的API设计与应用）。第13～15章是通信协议的实例（通信协议）。

第2～11章实验侧重于C语言基础，其中通过不同语法完成相同实验内容的方式，还在“卓越工程师培养系列”的其他语言类教材中得以体现，如Android、WinForm、MFC和Qt。在学习Java、C#、C++等编程语言时，同样是基于秒值-时间值转换实验。第12～15章实验侧重于应用，其中，第12章实验中的循环队列常常应用在串口通信中，如单片机（如STM32F1和STM32F4等）和DSP（如TMS320F28335等）的串口收发数据，在“卓越工程师培养系列”的单片机和DSP等教材中将会看到本书第12章所介绍的API函数的进一步应用。第13～15章是通信协议的API设计与应用，该协议基于主从机通信，从机作为执行单元，用于处理一些具体的事务，而主机（如Windows、Linux、Android和emWin平台等）常与从机进行交互，向从机发送命令，或处理来自从机的数据，在主机与从机的通信过程中，交互媒介就是通信协议，在“卓越工程师培养系列”的一系列涉及主从机通信的教材中，又可以看到这三章的影子。

本书的特点如下。

1. 以手把手的方式引导读者开展实验，通过实验原理介绍、实验步骤拆解和剖析，让读者快速入门；“本章任务”是实验的延伸和拓展，让读者通过实战巩固实验的知识点；“本章习题”用于检验读者是否掌握了书中的知识点。

2.“实验原理”详细讲解每个实验所涉及的知识点，未涉及的知识点基本不予介绍，以便于初学者快速掌握C语言程序设计的核心要点。

3．将C语言规范贯穿于整个程序设计过程中，如排版和注释规范、文件和函数命名规范，以及项目架构设计等。

4．配有丰富的资料包，包括C语言例程、软件包、PPT和视频等，资料包会持续更新，下载链接可通过微信公众号“卓越工程师培养系列”获取。

参与本书编写的有汪天富、董磊、郭文波、曾凡均、彭芷晴。电子工业出版社张小乐编辑为本书的出版做了大量的工作。特别感谢深圳大学生物医学工程学院、深圳市乐育科技有限公司和电子工业出版社的大力支持。在此一并致以衷心的感谢！

由于编者水平有限，书中难免有不成熟和错误的地方，恳请读者批评指正。读者反馈发现的问题、索取相关资料或遇实验平台技术问题，可发信至邮箱：ExcEngineer@163.com。

编　者

目　录

第 1 章　C 语言开发环境……1
1.1　C 语言开发环境简介……1
1.2　安装 Visual Studio Community 2019……1
1.2.1　计算机配置要求……1
1.2.2　安装软件……1
1.3　本书配套的资料包……5
本章任务……5
本章习题……5
第 2 章　简单的秒值-时间值转换……6
2.1　实验内容……6
2.2　实验原理……6
2.2.1　包含头文件……6
2.2.2　主函数……6
2.2.3　标识符与关键字……6
2.2.4　常用数据类型……7
2.2.5　常量和变量……7
2.2.6　局部变量命名规范……7
2.2.7　用 printf 函数输出数据……7
2.2.8　用 scanf_s 函数输入数据……8
2.2.9　算术运算符……8
2.2.10　程序注释……9
2.2.11　system("pause")的作用……9
2.3　实验步骤……9
本章任务……13
本章习题……13
第 3 章　基于数组的秒值-时间值转换……14
3.1　实验内容……14
3.2　实验原理……14
3.2.1　数组定义……14
3.2.2　数组的初始化……14
3.2.3　数组元素引用……14
3.2.4　数组命名规范……15
3.3　实验步骤……15
本章任务……16
本章习题……16
第 4 章　基于函数的秒值-时间值转换……17
4.1　实验内容……17
4.2　实验原理……17
4.2.1　为什么要使用函数……17
4.2.2　函数的定义……18
4.2.3　函数的声明……18
4.2.4　函数的调用……18
4.2.5　函数的参数……19
4.2.6　函数的返回值……19
4.2.7　内部函数……19
4.2.8　函数命名规范……19
4.3　实验步骤……19
本章任务……21
本章习题……21
第 5 章　基于枚举的秒值-时间值转换……22
5.1　实验内容……22
5.2　实验原理……22
5.2.1　用 typedef 声明新类型名……22
5.2.2　枚举类型……22
5.2.3　switch 语句……23
5.3　实验步骤……24
本章任务……26
本章习题……26
第 6 章　基于指针的秒值-时间值转换……27
6.1　实验内容……27
6.2　实验原理……27
6.2.1　大端模式和小端模式……27
6.2.2　常用的三种数据类型的存储方式……28
6.2.3　什么是指针……29
6.2.4　指针变量的定义和使用……30
6.2.5　指针变量注意事项……31
6.2.6　指针与数组……31
6.2.7　用 if 语句实现选择结构……32
6.2.8　逻辑运算符……33

6.2.9 程序调试 ······33
6.3 实验步骤 ······35
本章任务 ······36
本章习题 ······37
第 7 章 基于结构体的秒值-时间值转换 ······38
7.1 实验内容 ······38
7.2 实验原理 ······38
7.2.1 为什么要使用结构体类型 ······38
7.2.2 结构体类型的声明和使用方法 ······38
7.3 实验步骤 ······39
本章任务 ······40
本章习题 ······40
第 8 章 基于结构体指针的秒值-时间值转换 ······41
8.1 实验内容 ······41
8.2 实验原理 ······41
8.2.1 结构体指针 ······41
8.2.2 指针加 *M* 操作 ······41
8.3 实验步骤 ······42
本章任务 ······44
本章习题 ······44
第 9 章 基于多文件的秒值-时间值转换 ······45
9.1 实验内容 ······45
9.2 实验原理 ······45
9.2.1 项目架构 ······45
9.2.2 为什么要使用多文件 ······45
9.2.3 编译过程 ······46
9.2.4 防止重编译 ······46
9.2.5 API 函数 ······47
9.2.6 C 语言软件设计规范 ······48
9.3 实验步骤 ······48
本章任务 ······58
本章习题 ······58
第 10 章 基于多媒体定时器的电子钟设计 ······59
10.1 实验内容 ······59
10.2 实验原理 ······59
10.2.1 项目架构 ······59
10.2.2 函数指针 ······60
10.2.3 回调函数 ······60
10.2.4 局部变量和全局变量 ······60
10.2.5 静态变量 ······60
10.2.6 自增、自减运算符 ······60
10.2.7 多媒体定时器 ······60
10.3 实验步骤 ······61
本章任务 ······72
本章习题 ······72
第 11 章 电子钟的 API 设计与应用 ······73
11.1 实验内容 ······73
11.2 实验原理 ······73
11.2.1 项目架构 ······73
11.2.2 RunClock 模块函数 ······73
11.2.3 DataType.h ······75
11.2.4 while 循环语句 ······76
11.3 实验步骤 ······76
本章任务 ······85
本章习题 ······85
第 12 章 循环队列的 API 设计与应用 ······86
12.1 实验内容 ······86
12.2 实验原理 ······86
12.2.1 队列与循环队列 ······86
12.2.2 循环队列 Queue 模块函数 ······86
12.2.3 for 循环语句 ······89
12.3 实验步骤 ······90
本章任务 ······99
本章习题 ······99
第 13 章 协议处理的 API 设计与应用 ······100
13.1 实验内容 ······100
13.2 实验原理 ······100
13.2.1 PCT 通信协议 ······100
13.2.2 PCT 通信协议格式 ······101
13.2.3 PCT 通信协议打包过程 ······102
13.2.4 PCT 通信协议解包过程 ······103
13.3 实验步骤 ······104
本章任务 ······116
本章习题 ······116
第 14 章 模拟从机命令接收与数据发送 ······117
14.1 实验内容 ······117
14.2 实验原理 ······118
14.2.1 wave1 和 wave2 模块的命令包和数据包 ······118
14.2.2 新增 wave1 和 wave2 模块通信协议 ······119

14.2.3 从机命令接收流程说明………120
14.2.4 从机数据发送流程说明………120
14.3 实验步骤………121
本章任务………136
本章习题………137
第 15 章 模拟主机命令发送与数据接收……138
15.1 实验内容………138
15.2 实验原理………138
15.2.1 主机命令发送流程说明………138
15.2.2 主机数据接收流程说明………138
15.3 实验步骤………139
本章任务………151
本章习题………152
附录 A C 语言软件设计规范（LY-STD001-2019）………153
A.1 排版………153
A.2 注释………154
A.3 命名规范………154
A.4 C 文件模板………156
A.5 H 文件模板………158
参考文献………160

第1章　C语言开发环境

1.1　C语言开发环境简介

支持C语言的开发环境有很多，如VS Code、Sublime Text、Atom和Visual Studio等，其中，前3个实质上是轻便型代码编辑器，配置C语言编译器后，就可以直接作为C语言开发环境。Visual Studio是软件开发领域中非常有名的集成开发环境（Integrated Development Environment，IDE），全称为Microsoft Visual Studio，是微软公司的开发工具包系列产品。Visual Studio是一个基本完整的开发工具集，包括了整个软件生命周期中所需要的大部分工具，如UML工具、代码管控工具、集成开发环境等。

微软提供了诸多版本的Visual Studio，本书使用的是Visual Studio 2019的Community版本，可以在官网上免费获得，虽然该版本相对于付费版本缺少了某些功能，但并不影响学习本书内容。本书中的实验例程使用的版本为Visual Studio Community 2019。

1.2　安装Visual Studio Community 2019

1.2.1　计算机配置要求

在安装Visual Studio Community 2019之前，为了保证开发顺畅，建议选用配置较高的计算机。对计算机的配置要求如下。

（1）操作系统：Windows 7及以上版本（本书基于Windows 10，推荐使用Windows 10）；

（2）CPU：主频不低于2.0GHz；

（3）内存：4GB或更高，推荐8GB；

（4）硬盘：80GB或更大。

1.2.2　安装软件

很多初学者在安装软件时，由于操作系统差异、安装软件不当等原因，走了很多弯路，为避免上述情况，建议严格按照以下步骤进行软件安装。

本书用到的相关软件均放在配套资料包的“02.相关软件”文件夹中。在安装Visual Studio软件之前，建议先安装.NET Framework 4.6框架，可在“02.相关软件\.NET Framework 4.6”文件夹中双击运行NDP46-KB3045557-x86-x64-AllOS-ENU.exe进行安装。在安装过程中，如果系统弹出“这台计算机中已经安装了.NET Framework 4.6或版本更高的更新”的提示信息，则不必再安装.NET Framework 4.6框架了。注意，Visual Studio需处于联网状态下安装。使用Windows 7操作系统安装时，若遇到无法联网下载的情况，可以尝试安装“02.相关软件\补丁文件”文件夹中的两个补丁文件kb4490628和kb4474419来解决，双击运行即可开始安装。

双击运行本书配套资料包“02.相关软件\Visual Studio Community 2019”文件夹中的vs_community_408779306.1590572925.exe，在如图1-1所示的对话框中，单击“继续”按钮。

图 1-1　Visual Studio 安装步骤 1

系统弹出如图 1-2 所示的安装界面，等待准备就绪。

Visual Studio Installer

即将完成...一切即将准备就绪。

已下载

已安装

图 1-2　Visual Studio 安装步骤 2

在如图 1-3 所示的对话框中，在“工作负载”标签页下勾选“.NET 桌面开发”和“使用 C++的桌面开发”项，并在“可选”栏中勾选“适用于最新 v142 生成工具的 C++ MFC……”项。单击“安装”按钮。

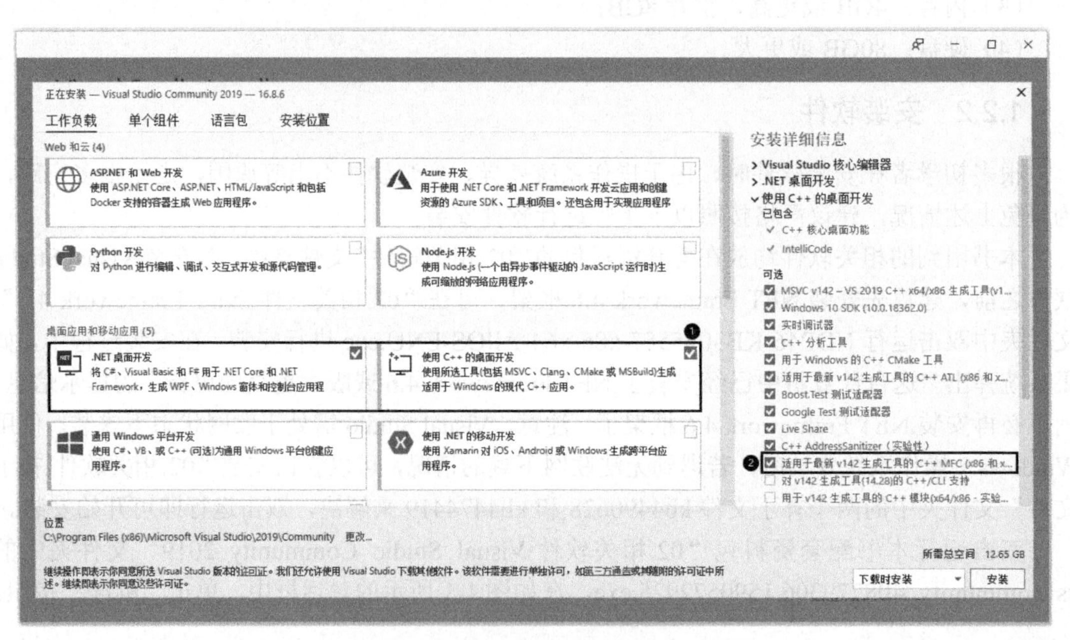

图 1-3　Visual Studio 安装步骤 3

如图 1-4 所示为安装界面。

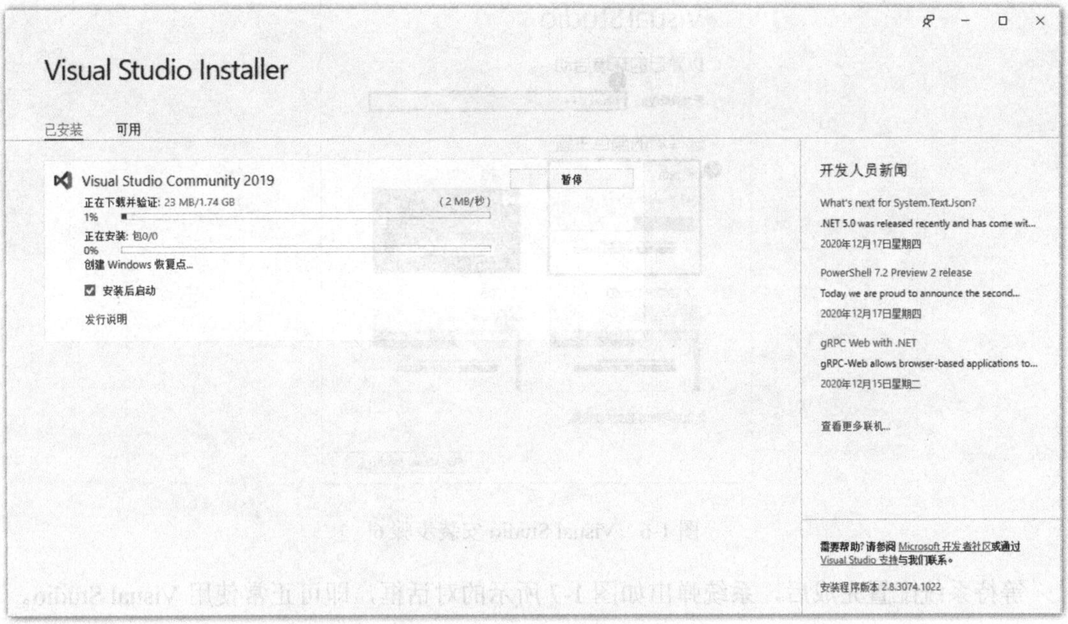

图 1-4　Visual Studio 安装步骤 4

安装完成后，系统弹出如图 1-5 所示的对话框。若已有账户，可以直接登录账户；若没有，可选择“以后再说”或“创建一个”按钮。

图 1-5　Visual Studio 安装步骤 5

如图 1-6 所示，在“开发设置”中选择“Visual C++”，选择合适的颜色主题后，单击“启动 Visual Studio”按钮。

图 1-6　Visual Studio 安装步骤 6

等待系统配置完成后，系统弹出如图 1-7 所示的对话框，即可正常使用 Visual Studio。

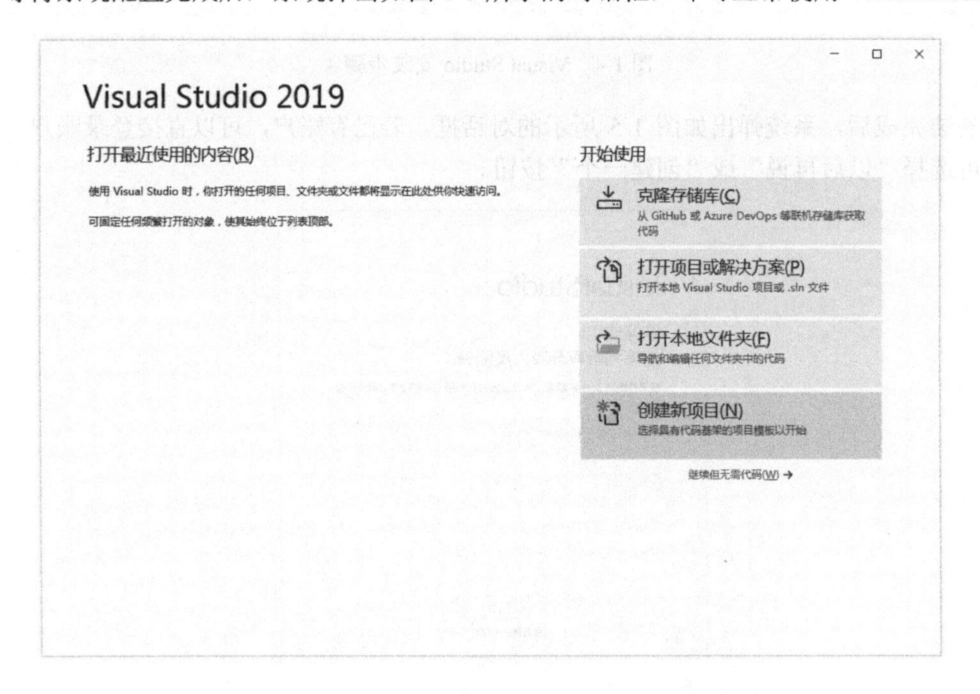

图 1-7　Visual Studio 安装步骤 7

程序块通常采用缩进风格编写，本书建议缩进 2 个空格。同时将 Tab 键设置为 2 个空格，这样可以防止使用不同的编辑器阅读代码时出现代码布局不整齐的现象。注意，对于由开发工具自动生成的代码可以不一致。针对 Visual Studio Community 2019 软件，设置制表符长度和缩进长度的具体方法如图 1-8 所示，①在 Visual Studio Community 2019 软件中，执行菜单命令“工具”→“选项”；②在弹出的“选项”对话框中，单击“文本编辑器”→“C/C++”→“制表符”，选择“插入空格”；③“制表符大小”和“缩进大小”均设置为 2，即可完成制表符长度和缩进长度设置。

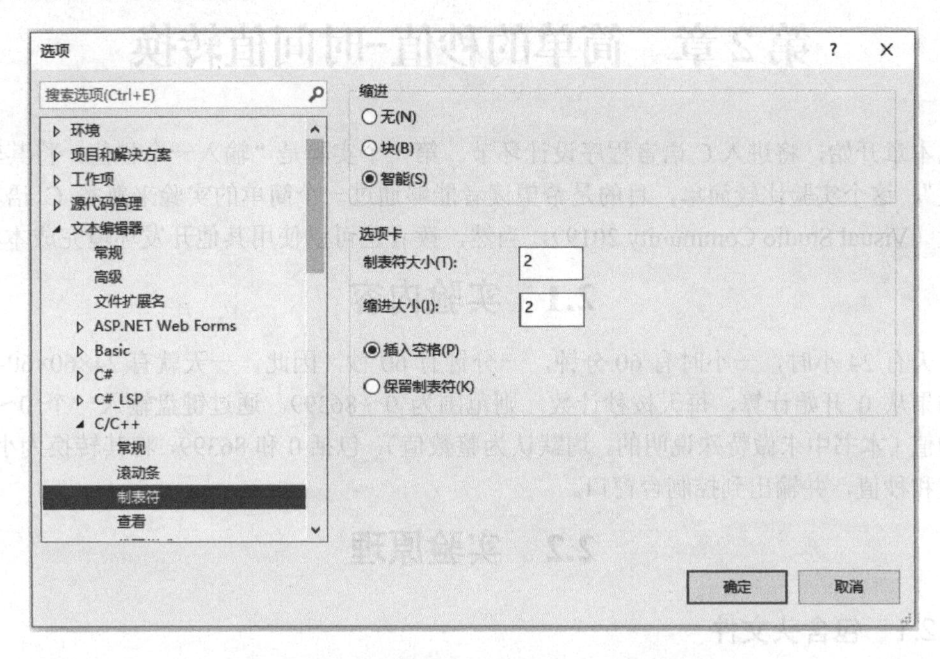

图1-8　Visual Studio Community 2019软件设置

1.3　本书配套的资料包

本书配套的"C语言程序设计资料包"可通过微信公众号"卓越工程师培养系列"提供的链接获取。为了保持与本书实践操作的一致性，建议将资料包复制到计算机的D盘中，地址即为"D:\C语言程序设计资料包"。资料包由若干文件夹组成，如表1-1所示。

表1-1　C语言程序设计资料包清单

序号	文件夹名	文件夹介绍
1	C语言入门资料	存放学习C语言程序设计相关的入门资料，建议读者在开始实验前，先阅读C语言入门资料
2	相关软件	存放本书使用到的软件，如Visual Studio Community 2019
3	例程资料	存放C语言程序设计所有实验的相关素材，读者根据这些素材开展各个实验
4	PPT讲义	存放每个章节的PPT讲义
5	视频资料	存放配套视频资料
6	软件资料	存放本书使用到的小工具，如PCT打包解包工具

本章任务

学习完本章后，下载本书配套的资料包，并完成Visual Studio Community 2019的安装。

本章习题

1．C语言都有哪些开发环境？

2．Microsoft Visual Studio软件都有哪些功能？

第 2 章　简单的秒值-时间值转换

从本章开始，将进入 C 语言程序设计环节。第一个实验是“输入一个秒值，将其转换为时间值”，这个实验比较简单，目的是希望读者能够通过一个简单的实验来熟悉 C 语言的开发环境（Visual Studio Community 2019），当然，读者也可以使用其他开发环境完成本实验。

2.1　实验内容

一天有 24 小时，一小时有 60 分钟，一分钟有 60 秒，因此，一天就有 24×60×60=86400 秒，如果从 0 开始计算，每天按秒计数，则范围为 0～86399。通过键盘输入一个 0～86399 之间的值（本书中未做特殊说明的，均默认为整数值），包括 0 和 86399，将其转换为小时值、分钟值和秒值，并输出到控制台窗口。

2.2　实验原理

2.2.1　包含头文件

头文件是包含了函数声明、宏定义和枚举结构体定义等的一个文件，头文件分为系统自带的和用户编写的。包含头文件是一条预处理指令，它的处理过程是通过预处理器读入源代码，根据预处理指令对源程序进行替换，然后再交给编译器。

包含头文件有两种方式，一种是尖括号包含（如#include <stdio.h>），另一种是双引号包含（如#include "UserFile.h"）。注意，尖括号包含通常用于包含标准库的头文件，编译器会去系统配置的库环境变量或用户配置的路径中搜索，而不会在项目的当前目录中查找；双引号包含通常用于包含用户编写的头文件，编译器会先在项目的当前目录中查找，找不到后才会去系统配置的库环境变量和用户配置的路径中搜索。

2.2.2　主函数

C 语言程序是由函数构成的，主函数是其中最核心的部分。一个 C 语言程序有且仅有一个主函数，即 main 函数，这是程序的唯一入口。

2.2.3　标识符与关键字

标识符就是给常量、变量、数组和函数等定义的名称，其命名需要遵循一定的规则：

（1）必须由字母、数字或下画线组成；

（2）不能以数字开头；

（3）不能是关键字。

在 C 语言中，有一些单词被赋予了特定的意义，即关键字。标识符的命名应避开关键字，ANSI C 标准 C 语言中的关键字有 32 个，如表 2-1 所示。

表 2-1　C 语言中的关键字

auto	break	case	char	const	continue	default	do
double	else	enum	extern	float	for	goto	while

续表

int	long	register	return	short	signed	sizeof	static
struct	switch	typedef	union	unsigned	void	volatile	if

2.2.4　常用数据类型

本书中常用的数据类型主要有 int 型、short 型和 char 型，如表 2-2 所示。

表 2-2　常用的数据类型

类　型	关 键 字	字 节 数	取 值 范 围
基本整型	int	4	$-2^{31}\sim(2^{31}-1)$
无符号基本整型	unsigned int	4	$0\sim(2^{32}-1)$
短整型	short	2	$-2^{15}\sim(2^{15}-1)$
无符号短整型	unsigned short	2	$0\sim(2^{16}-1)$
字符型	char	1	$-2^{7}\sim(2^{7}-1)$
无符号字符型	unsigned char	1	$0\sim(2^{8}-1)$

2.2.5　常量和变量

在程序运行过程中，值不能被改变的量称为常量，而变量的值是可以改变的。变量必须先定义后使用，通常在定义变量时直接为其赋初值，一般形式如下：

```
类型名 变量名 = 常数/表达式;
```

类型名指定变量的数据类型，“=”为赋值符号，将右边的常数或表达式的值赋给变量。

2.2.6　局部变量命名规范

函数内部的非静态变量为局部变量，其有效区域仅限于函数范围内，局部变量命名采用第一个单词首字母小写，后续单词的首字母大写，其余字母小写格式，如 timerStatus、tickVal、restTime。

2.2.7　用 printf 函数输出数据

在使用 C 语言进行程序设计时，经常需要打印一些提示信息，printf 函数具有打印功能，printf 函数的一般格式如下：

```
printf(格式控制, 输出表列);
```

格式控制包含格式声明和普通字符两部分，格式声明由“%”和格式字符组成，其作用是将输出数据转换为指定的格式输出，普通字符则是需要保持原样输出的字符。输出表列是需要输出的数据，可以是常量、变量或表达式。

例如，执行以下语句：

```
printf("Output: %d%d", a, b);
```

如果 a=10，b=20，则打印出以下信息：

```
Output: 10 20
```

printf 函数中常用的格式字符如表 2-3 所示。

表 2-3　printf 函数中常用的格式字符

格式字符	说　明
d, i	以带符号的十进制形式输出整数（正数不输出符号）
o	以八进制无符号形式输出整数
x, X	以十六进制无符号形式输出整数，若用 x，则在输出十六进制的 a～f 时以小写形式输出；若用 X，则以大写形式输出
c	以字符形式输出，只输出一个字符
s	输出字符串
f	以小数形式输出

2.2.8　用 scanf_s 函数输入数据

printf 函数用于打印提示信息，用户可以使用 scanf_s 函数输入数据，scanf_s 函数的一般格式如下：

```
scanf_s (格式控制, 地址表列);
```

格式控制的含义同 printf 函数。地址表列是由若干变量的地址组成的表列，表示变量地址的方法是在变量名前加取地址符“&”。

例如，执行以下语句：

```
scanf("%d", &val);
```

如果通过键盘输入 10，则 val=10。

scanf_s 函数中常用的格式字符的用法与 printf 函数类似，如表 2-4 所示。

表 2-4　scanf_s 函数中常用的格式字符

格式字符	说　明
d,i	输入有符号的十进制整数
o	输入无符号的八进制整数
x,X	输入无符号的十六进制整数，大小写作用相同
c	输入单个字符
s	输入字符串
f	输入实数，可以为小数形式或指数形式

2.2.9　算术运算符

常用的算术运算符如表 2-5 所示。

表 2-5　常用的算术运算符

运算符	含　义	示　例	结　果
+	取正值	+a	a 的值
−	取负值	−a	a 的算术负值

续表

运 算 符	含 义	示 例	结 果
+	加法	a+b	a 与 b 的和
–	减法	a–b	a 与 b 的差
*	乘法	a*b	a 与 b 的积
/	除法	a/b	a 与 b 的商
%	取余	a%b	a 除以 b 的余数

2.2.10　程序注释

程序注释即为对程序代码的解释说明，可以增加代码的可读性。一段好的代码，注释是必不可少的。常见的注释形式有两种：单行注释和多行注释。单行注释使用“//”符号，作用范围是从“//”开始到本行结束；多行注释使用“/*　*/”符号，作用范围是“/*”和“*/”之间。

2.2.11　system("pause")的作用

如果代码中没有 system("pause")，Windows 控制台程序的结果输出界面就会一闪而过，基本上看不到执行结果。添加 system("pause")之后，系统会在 Windows 控制台程序结果输出界面的最后一行输出“请按继续键继续...”，等待用户按下一个按键，然后才会退出结果输出界面，这样用户就可以看清执行结果。

2.3　实验步骤

首先，在计算机的 D 盘中建立一个 CProgramTest 文件夹，将本书配套资料包的“03.例程资料\Material\01.输入一个秒值将其转换为时间值实验”文件夹复制到 CProgramTest 文件夹中。当然，项目保存的文件夹路径可自行选择，但是完整项目保存的文件夹及命名一定要严格按照要求进行设置，从细微之处养成良好的规范习惯。

然后，打开 Visual Studio Community 2019 软件，单击“创建新项目”按钮，如图 2-1 所示。

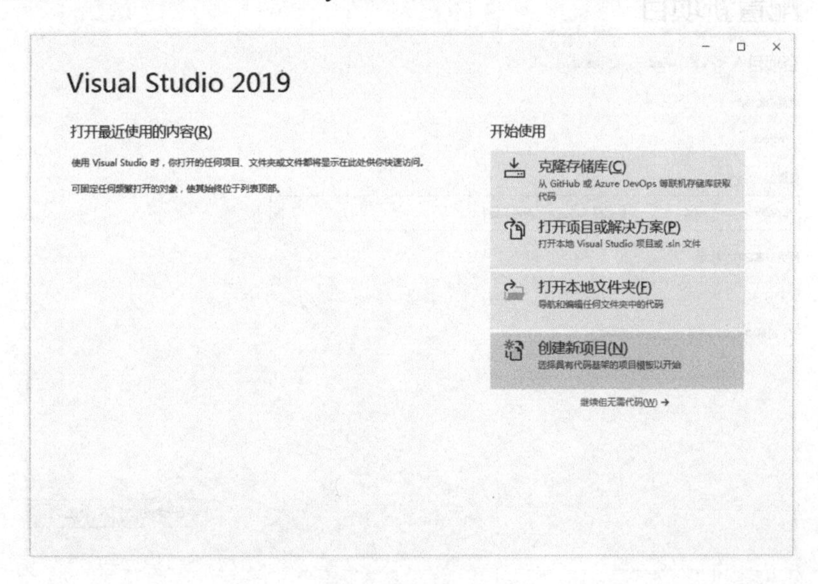

图 2-1　创建新项目

本书涉及的例程不需要使用复杂的窗体交互，均使用控制台应用程序，因此，选择创建的项目类型时，单击“空项目”按钮，然后单击“下一步”按钮，如图 2-2 所示。

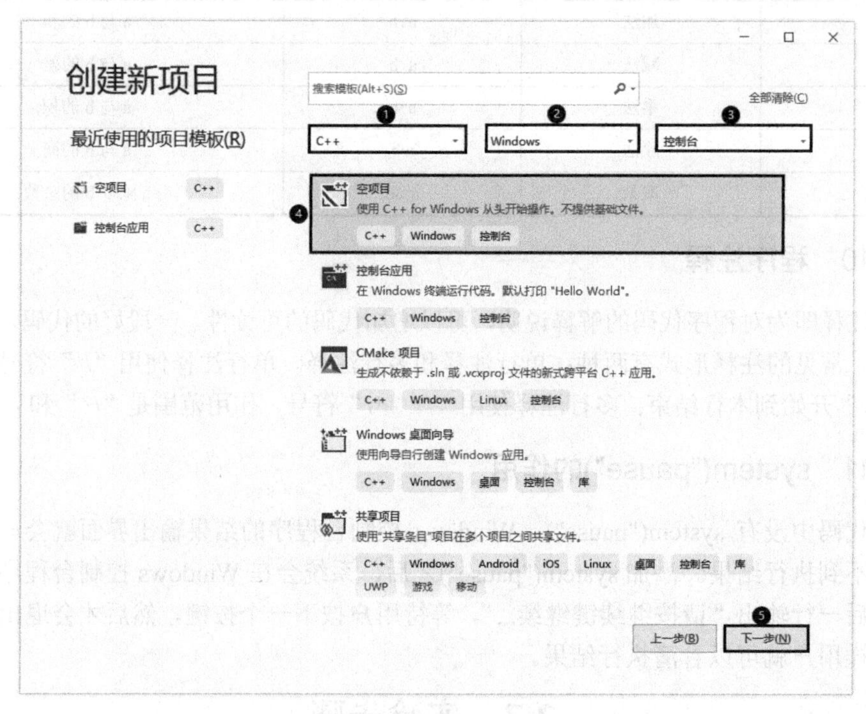

图 2-2　选择项目模板

如图 2-3 所示，设置项目名称为 Project，位置选择“D:\CProgramTest\01.输入一个秒值将其转换为时间值实验\”文件夹，勾选“将解决方案和项目放在同一目录中”项，然后单击“创建”按钮。

图 2-3　配置新项目

项目创建成功后，新建 C 文件，右键单击项目名“Project”，在快捷菜单中选择“添加”→“新建项”，如图 2-4 所示。

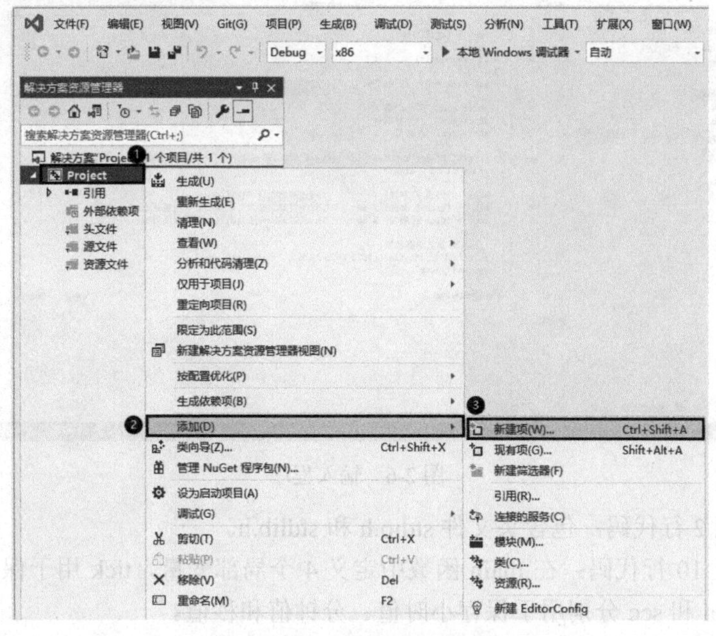

图 2-4　添加新建项

在弹出的如图 2-5 所示的“添加新项-Project”对话框中，进行以下操作：①单击“C++文件(.cpp)”；②名称设置为 App.c；③位置选择“D:\CProgramTest\01.输入一个秒值将其转换为时间值实验\Code\”；④单击“添加”按钮，这样就完成了在项目下添加 App.c 文件的操作。注意，文件名是 App.c，而不是 App.cpp；存放 C 文件的文件夹是 Code，而不是 Project。

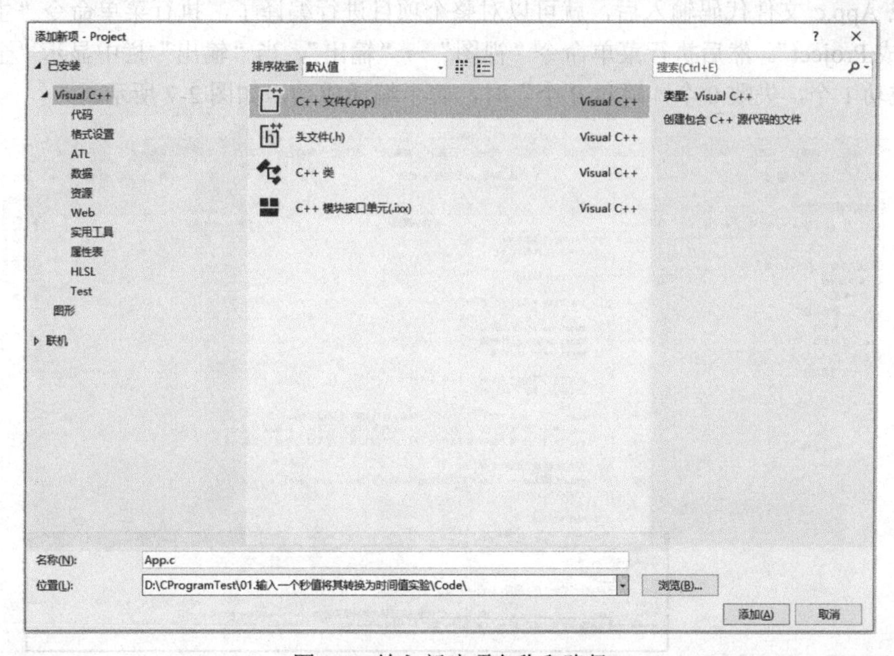

图 2-5　输入新建项名称和路径

在 App.c 文件中输入如图 2-6 所示编程界面中的代码。下面按照顺序对部分语句进行解释。

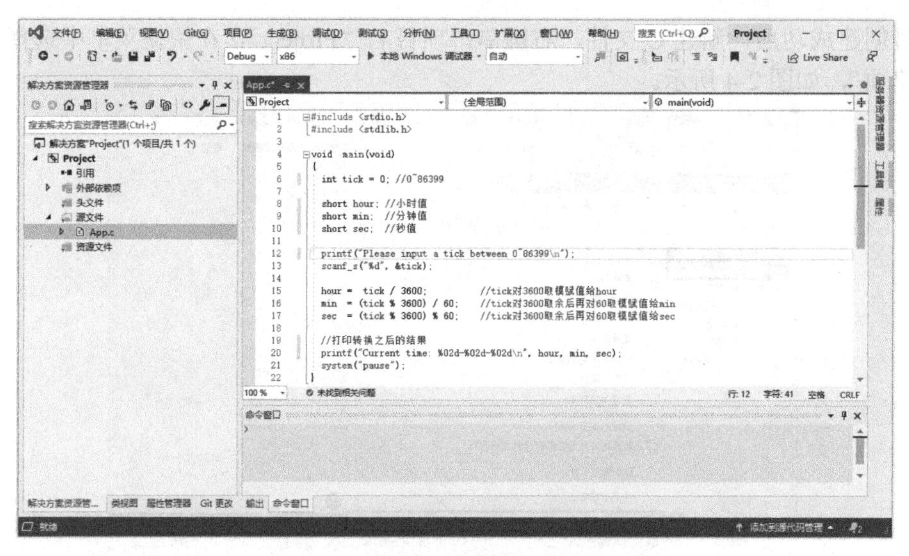

图 2-6　输入程序

（1）第 1 至 2 行代码：包含头文件 stdio.h 和 stdlib.h。

（2）第 6 至 10 行代码：在 main 函数中定义 4 个局部变量，tick 用于保存时间值对应的秒值，hour、min 和 sec 分别用于保存小时值、分钟值和秒值。

（3）第 12 至 13 行代码：通过 printf 函数打印提示信息，提示用户输入一个 0～86399 之间的值，然后，通过 scanf_s 函数获取键盘输入的内容，并转换为 32 位整型数据赋值给 tick。

（4）第 15 至 17 行代码：将 tick 依次转换为小时值、分钟值和秒值。

（5）第 20 至 21 行代码：通过 printf 函数打印转换之后的时间结果，格式为“小时-分钟-秒”。system 函数用于解决输出界面一闪而过的问题。

完成 App.c 文件代码输入后，就可以对整个项目进行编译了，执行菜单命令“生成”→“重新生成 Project”。然后执行菜单命令“视图”→“输出”，当“输出”栏中显示“全部重新生成：成功 1 个，失败 0 个，跳过 0 个”时，表示编译成功，如图 2-7 所示。

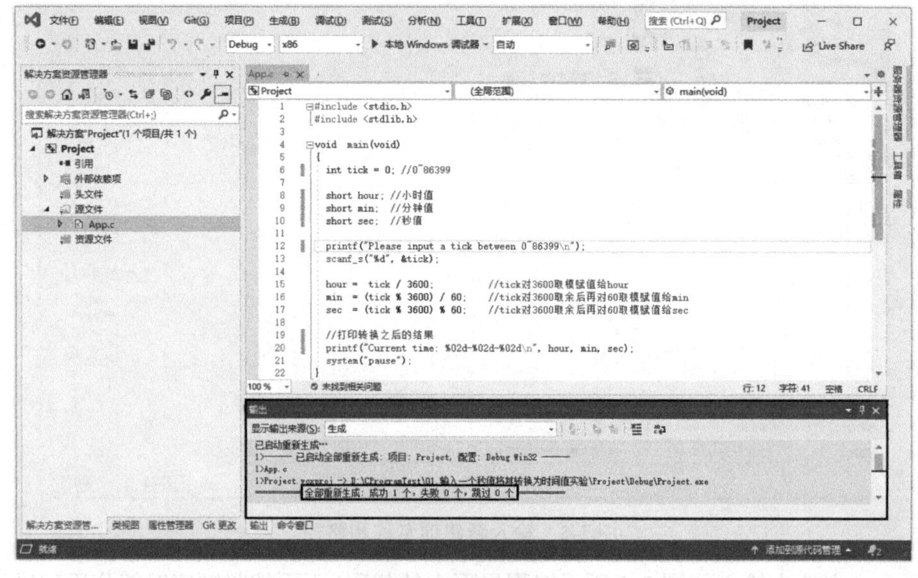

图 2-7　编译项目

最后，按 F5 键编译并运行程序，在弹出的控制台窗口中，输入 80000 后按回车键，可以看到运行结果，即输出“Current time: 22-13-20”，说明实验成功，如图 2-8 所示。

图 2-8　本章实验运行结果

本 章 任 务

任务 1：2020 年有 366 天，将 2020 年 1 月 1 日作为计数起点，即计数 1，2020 年 12 月 31 日作为计数终点，即计数 366。计数 1 代表“2020 年 1 月 1 日-星期三”，计数 10 代表“2020 年 1 月 10 日-星期五”。参考本章实验，通过键盘输入一个 1～366 之间的值，包括 1 和 366，将其转换为年、月、日、星期，并输出到控制台窗口。

任务 2：参考本章实验，通过键盘输入 2 个数，判断大小，并输出结果到控制台窗口。

任务 3：参考本章实验，通过键盘输入 10 个数，判断大小，并输出结果到控制台窗口。

本 章 习 题

1．简述基于 Visual Studio Community 2019 开发环境的 C 语言程序设计流程。

2．scanf 和 scanf_s 有什么区别？

3．为什么要设置制表符长度和缩进长度？

4．包含头文件有两种方式，一种是尖括号包含，另一种是双引号包含，简述两种包含头文件方式的区别。

5．局部变量的有效区域是什么？本书中，局部变量命令采用的是什么格式？

6．如果不使用 system("pause")语句，还有什么方法可以避免 Windows 控制台结果输出界面一闪而过的现象。

第 3 章　基于数组的秒值-时间值转换

数组是用来存储一系列数据的，但它往往被认为是一系列相同类型的变量。本章的实验是基于数组实现秒值-时间值转换，通过该实验，熟悉数组的命名、定义、元素引用和初始化。

3.1　实验内容

通过键盘输入一个 0～86399 之间的值，包括 0 和 86399，将其转换为小时值、分钟值和秒值，而小时值、分钟值和秒值为数组 arrTimeVal 的元素，即 arrTimeVal[2]为小时值、arrTimeVal[1]为分钟值、arrTimeVal[0]为秒值，并将结果输出到控制台窗口。

3.2　实验原理

3.2.1　数组定义

数组是相同类型数据的有序集合，定义数组的一般形式如下：

```
类型说明符 数组名 [常量表达式];
```

其中，类型说明符是任意一种基本数据类型或构造数据类型，数组名是用户定义的数组标识符，方括号中的常量表达式表示数据元素的个数，也称为数组的长度。例如：

```
int arrCircleBuf[10];
```

int 表示数组元素的类型为整型，arrCircleBuf 为数组名，方括号中的 10 表示数组中包含 10 个元素，从 arrCircleBuf[0] 到 arrCircleBuf[9]。

3.2.2　数组的初始化

定义数组之后要先进行初始化才能使用，通常在定义时直接对数组元素赋初值，例如：

```
int arrTimeVal[2] = {512, 1024};
```

执行完上述语句，就实现了将 512 赋值给 arrTimeVal[0]，将 1024 赋值给 arrTimeVal[1]。

3.2.3　数组元素引用

数组元素表示的一般形式如下：

```
数组名[下标];
```

例如，引用上述代码中已经初始化的数组 arrTimeVal[1]，即数组中的第 2 个元素，如下所示：

```
arrTimeVal[1] = 3;//将 3 赋值给 arrTimeVal[1]
int dispVal = 0;//定义并初始化整型变量 dispVal
dispVal = arrTimeVal[1];//将 arrTimeVal[1]赋值给 dispVal
```

3.2.4 数组命名规范

局部变量命名规范也适用于函数内的非静态数组命名：第一个单词的首字母小写，后续单词的首字母大写，其余字母小写。但本书建议在数组名前加 arr 前缀，以区别于其他变量，如 arrSendData、arrRestTime、arrTempData。

3.3 实验步骤

首先，将本书配套资料包的“03.例程资料\Material\02.基于数组的秒值-时间值转换实验”文件夹复制到 CProgramTest 文件夹中，然后，双击运行“D:\CProgramTest\02.基于数组的秒值-时间值转换实验\Project”文件夹中的 Project.sln 文件，最后，将程序清单 3-1 中的代码输入 App.c 文件中。下面按照顺序对部分语句进行解释。

（1）第 8 行代码：声明一个 short 型数组，数组名为 arrTimeVal，可以存放 3 个 short 型数据。

（2）第 13 至 15 行代码：通过 tick 计算小时值、分钟值和秒值，分别赋值给 arrTimeVal[2]、arrTimeVal[1]、arrTimeVal[0]。

（3）第 17 行代码：通过 printf 函数打印转换之后的时间结果，时间分别通过数组的引用获得，格式为“Current time: 小时-分钟-秒”。

程序清单 3-1

```
1.   #include <stdio.h>
2.   #include <stdlib.h>
3.
4.   void  main(void)
5.   {
6.     int tick = 0; //0~86399
7.
8.     short arrTimeVal[3]; //存放小时值、分钟值、秒值
9.
10.    printf("Please input a tick between 0~86399\n");
11.    scanf_s("%d", &tick);
12.
13.    arrTimeVal[2] =  tick / 3600;         //tick 对 3600 取模赋值给 arrTimeVal[2] ，即小时值
14.    arrTimeVal[1] = (tick % 3600) / 60; //tick 对 3600 取余后再对 60 取模赋值给 arrTimeVal[1],
                                                                                    即分钟值
15.    arrTimeVal[0] = (tick % 3600) % 60; //tick 对 3600 取余后再对 60 取余赋值给 arrTimeVal[0],
                                                                                    即秒值
16.
17.    printf("Current time: %02d-%02d-%02d\n", arrTimeVal[2], arrTimeVal[1], arrTimeVal[0]);
18.
19.    system("pause");
20.  }
```

最后，按 F5 键编译并运行程序，在弹出的控制台窗口中，输入 80000 后按回车键，可以看到运行结果，即输出“Current time: 22-13-20”，说明实验成功。

本 章 任 务

任务1：2020年有366天，将2020年1月1日作为计数起点，即计数1，2020年12月31日作为计数终点，即计数366。计数1代表“2020年1月1日-星期三”，计数10代表“2020年1月10日-星期五”。参考本章实验，通过键盘输入一个1～366之间的值，包括1和366，基于数组，将其转换为年、月、日、星期，并输出到控制台窗口。

任务2：参考本章实验，通过键盘输入2个数，判断大小，并输出结果到控制台窗口。

任务3：参考本章实验，通过键盘输入10个数，判断大小，并输出结果到控制台窗口。

本 章 习 题

1．如何定义数组？

2．如何引用数组元素？

3．如何初始化数组？

4．本书中，数组命名采用的是什么格式？

第 4 章　基于函数的秒值-时间值转换

函数是一组共同执行一个任务的语句，每个 C 语言程序都至少有一个函数，即主函数 main，用户也可以自定义其他函数。本章的实验是基于函数实现秒值-时间值转换，通过该实验，熟悉函数的声明、定义和调用等方法。

4.1　实验内容

通过键盘输入一个 0～86399 之间的值，包括 0 和 86399，用 CalcHour 函数计算小时值，用 CalcMin 函数计算分钟值，用 CalcSec 函数计算秒值，在主函数中通过调用上述三个函数实现秒值-时间值转换，并将结果输出到控制台窗口。

4.2　实验原理

4.2.1　为什么要使用函数

总经理作为一个公司的主要负责人，不可能直接管理每一位员工。通常，公司会设定若干部门，部门下又设定若干项目组，项目组下面才是每一位员工。总经理只需管理好各部门经理，部门经理负责管理项目组长，项目组长负责管理每一位员工，管理架构如图 4-1 所示。这种层次清晰的管理架构会让公司变得更加规范和高效。

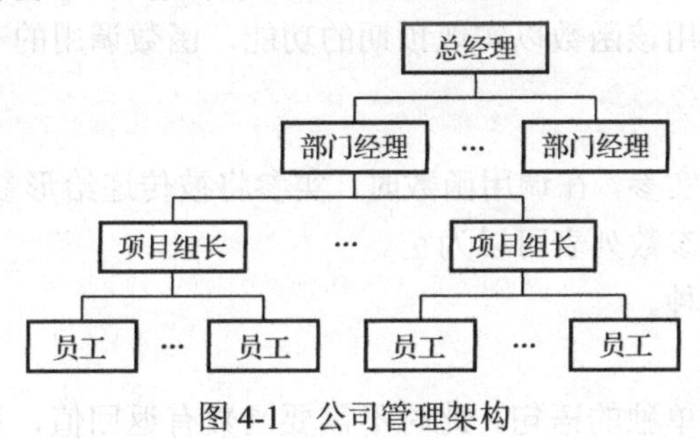

图 4-1　公司管理架构

程序设计也是如此，主函数（main 函数）相当于总经理，子函数相当于部门经理，或项目组长，或员工。因此，不建议将所有的代码都写在主函数中，正确的做法是将主函数中的某些功能模块包装成为一个子函数，主函数只需调用这些子函数即可，具体的功能均在子函数中实现。当然，子函数也可以调用更低一级的子函数，如孙函数和曾孙函数等，如图 4-2 所示。这种函数调用体系会让整个项目开发变得更加规范和清晰。

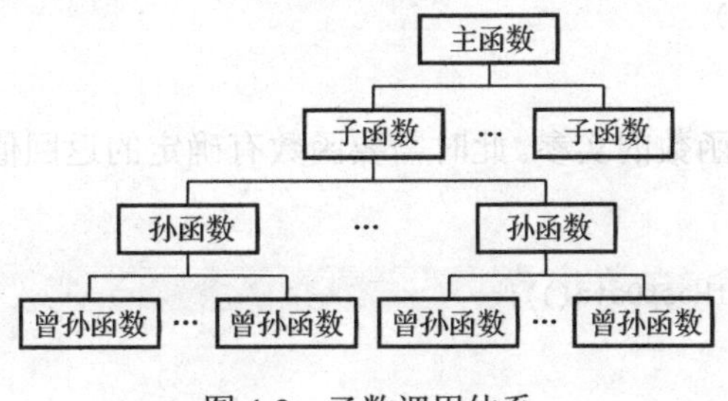

图 4-2　函数调用体系

4.2.2　函数的定义

与变量、数组一样，在使用函数前需要先定义函数，定义函数的一般格式如下：

```
返回值类型名 函数名(参数列表)
{
    函数体;
}
```

其中，返回值类型名即类型标识符，用来指定该函数返回值的类型。若没有返回值，则类型名为 void。参数列表是带有数据类型的变量名列表，称为形参，参数之间用逗号隔开。若函数没有参数，参数列表可以为 void 或为空。函数体包含声明部分和语句部分，是实现功能的主体。函数体可以为空，此时调用该函数没有任何实际作用。

4.2.3　函数的声明

自定义函数即开发者自己定义的函数，编译器不知道此函数的存在，因此在使用此函数之前需要先告知编译器，这个过程称为函数的声明。声明的作用是把函数名、函数参数的个数和参数类型等信息通知编译器，以便在遇到函数调用时，编译器能正确识别函数并检查调用是否合法。函数声明的一般格式如下：

```
返回值类型名 函数名(参数列表);
```

4.2.4　函数的调用

定义函数的目的是调用该函数以实现预期的功能，函数调用的一般形式如下：

```
函数名(参数列表);
```

此处的参数列表称为实参，在调用函数时，实参将被传递给形参，参数之间用逗号隔开。若调用的是无参函数，则参数列表可以为空。

函数调用的方式有三种。

（1）函数调用语句。

把函数调用作为一个单独的语句。此时不需要函数有返回值，只要求函数实现相应的功能。例如：

```
UnPackData(recData)
```

（2）函数表达式。

函数出现在一个表达式中。此时需要函数有确定的返回值参与表达式的运算。例如：

```
valid = PackData(pPackSent);
```

（3）函数参数。

函数调用作为另一个函数的实参。此时需要函数有确定的返回值作为另一个函数的实参。例如：

```
SendDataToMCU(dataType, GetHostData());
```

4.2.5　函数的参数

前文已经提到，在定义函数时括号里的参数列表为形参，在调用函数时括号里的参数列表为实参。在调用函数的过程中，系统会把实参的值传递给形参，从而参与函数的运算。

4.2.6　函数的返回值

通常，希望通过调用函数使主调函数得到一个确定的值，这就是函数值，也称为函数的返回值。函数的返回值是通过函数体中的 return 语句获得的。

在函数定义时指定了函数的返回值类型，return 语句的返回值类型应与函数的返回值类型一致，即函数的返回值类型决定函数体中返回值的类型。

4.2.7　内部函数

如果一个函数只能被同文件中的其他函数所调用，则称为内部函数，也称为内部静态函数。声明内部函数的一般格式如下：

```
static 类型名 函数名(形参列表);
```

例如：

```
static int Adder(int a, int b);
```

定义内部函数的一般格式如下：

```
static int Adder(int a, int b)
{
    int sum;

    sum = a + b;

    return(sum);
}
```

本书建议，内部函数必须加 static 关键字，在定义前必须先声明，且内部函数的声明与定义放在同一个文件中，声明完再逐个定义内部函数。注意，与内部函数对应的是 API 函数，API 函数将在后续章节中介绍。

4.2.8　函数命名规范

函数的命名可采用“动词+名词”的形式，关键部分建议采用完整的单词，辅助部分可采用缩写，缩写应符合英文的规范，每个单词的首字母大写，如 AnalyzeSignal、SendDataToPC、ReadBuffer。

4.3　实验步骤

首先，将本书配套资料包的“03.例程资料\Material\03.基于函数的秒值-时间值转换实验”文件夹复制到 CProgramTest 文件夹中，然后，双击运行“D:\CProgramTest\03.基于函数的秒值-时间值转换实验\Project”文件夹中的 Project.sln 文件，最后，将程序清单 4-1 中的代码输入 App.c 文件中。下面按照顺序对部分语句进行解释。

（1）第 4 至 39 行代码：声明和定义计算小时值的 CalcHour 函数、计算分钟值的 CalcMin 函数和计算秒值的 CalcSec 函数。

（2）第 53 至 55 行代码：通过调用 CalcHour、CalcMin 和 CalcSec 函数分别计算小时值、分钟值和秒值。

程序清单 4-1

```
1.  #include <stdio.h>
2.  #include <stdlib.h>
3.
4.  //内部函数 CalcHour 声明
5.  static  short CalcHour(int tick);
6.  //内部函数 CalcMin 声明
7.  static  short CalcMin(int tick);
8.  //内部函数 CalcSec 声明
9.  static  short CalcSec(int tick);
10.
11. //计算小时的内部函数实现
12. static  short CalcHour(int tick)
13. {
14.   short hour;
15.
16.   hour = tick / 3600;            //tick 对 3600 取模，赋值给 hour
17.
18.   return(hour);
19. }
20.
21. //计算分钟的内部函数实现
22. static  short CalcMin(int tick)
23. {
24.   short min;
25.
26.   min = (tick % 3600) / 60;      //tick 对 3600 取余后再对 60 取模，赋值给 min
27.
28.   return(min);
29. }
30.
31. //计算秒的内部函数实现
32. static  short CalcSec(int tick)
33. {
34.   short sec;
35.
36.   sec = (tick % 3600) % 60;      //tick 对 3600 取余后再对 60 取余，赋值给 sec
37.
38.   return(sec);
39. }
40.
41. //主函数
42. void  main(void)
43. {
44.   int tick = 0;
45.
```

```
46.     short hour;
47.     short min;
48.     short sec;
49.
50.     printf("Please input a tick between 0~86399\n");
51.     scanf_s("%d", &tick);
52.
53.     hour = CalcHour(tick);
54.     min  = CalcMin(tick);
55.     sec  = CalcSec(tick);
56.
57.     printf("Current time: %02d-%02d-%02d\n", hour, min, sec);
58.
59.     system("pause");
60. }
```

最后，按 F5 键编译并运行程序，在弹出的控制台窗口中，输入 80000 后按回车键，可以看到运行结果，即输出“Current time: 22-13-20”，说明实验成功。

本 章 任 务

任务 1：2020 年有 366 天，将 2020 年 1 月 1 日作为计数起点，即计数 1，2020 年 12 月 31 日作为计数终点，即计数 366。计数 1 代表“2020 年 1 月 1 日-星期三”，计数 10 代表“2020 年 1 月 10 日-星期五”。参考本章实验，通过键盘输入一个 1～366 之间的值，包括 1 和 366，基于函数，将其转换为年、月、日、星期，并输出到控制台窗口。

任务 2：参考本章实验，通过键盘输入 2 个数，判断大小，并输出结果到控制台窗口。

任务 3：参考本章实验，通过键盘输入 10 个数，判断大小，并输出结果到控制台窗口。

本 章 习 题

1．为什么要使用函数？

2．什么是内部函数？

第5章　基于枚举的秒值-时间值转换

枚举是把一个变量可能的值一一列举出来，且变量的值只限于列举出来的值的范围。本章的实验是基于枚举实现秒值-时间值转换，通过该实验，熟悉枚举的定义和使用方法。

5.1　实验内容

通过键盘输入一个0～86399之间的值，包括0和86399，使用CalcTimeVal函数计算时间值（包括小时值、分钟值和秒值），通过枚举区分具体是哪一种时间值，返回值为是否计算成功标志，在主函数中通过调用CalcTimeVal函数实现秒值-时间值转换，并将结果输出到控制台窗口。

5.2　实验原理

5.2.1　用typedef声明新类型名

除了可以直接使用C语言提供的标准类型名（如int、short等），还可以用typedef指定新的类型名来代替已有的类型名，例如：

```
typedef int i32;
typedef unisigned int u32;
```

这样就可以用i32来定义变量，例如：

```
i32 hour, min, sec;
```

将hour、min和sec定义为i32类型，而i32等价于int，因此hour、min和sec都为基本整型变量。

5.2.2　枚举类型

如果一个变量只有几种可能的值，则可以定义为枚举类型。枚举类型的声明和使用方法有三种。

（1）方法一

```
enum EnumWeekDay{
MON = 0,
TUE,
WED,
THU,
FRI,
SAT,
SUN};      //声明枚举类型

EnumWeekDay workDay, weekEnd;     //定义枚举变量

//使用枚举变量
```

```
workDay = MON;
weekEnd = SUN;
```

（2）方法二

```
enum{
MON = 0,
TUE,
WED,
THU,
FRI,
SAT,
SUN} workDay, weekEnd;//声明、定义
//使用
workDay = MON;
weekEnd = SUN;
```

（3）方法三

```
typedef enum{
MON = 0,
TUE,
WED,
THU,
FRI,
SAT,
SUN} EnumWeekDay; //声明
EnumWeekDay workDay, weekEnd;        //定义
//使用
workDay = MON;
weekEnd = SUN;
```

本书建议采用第三种声明和使用方法。

5.2.3　switch 语句

switch 语句是一种多分支选择语句，其一般形式如下：

```
switch(表达式)
{
    case 常量 1 :
        语句 1;
    break;
    case 常量 2 :
        语句 2;
    break;
    …
    case 常量 n :
        语句 n;
    break;
    default:
        语句 n+1;
    break;
}
```

switch 语句的作用是根据表达式的值，跳转到不同的语句执行。当表达式的值与其中一个 case 标号中的常量相符时，就执行该 case 标号后面的语句，直至执行到“break;”语句跳出 switch 结构为止。若表达式的值与所有 case 标号的常量都不相符，则执行 default 标号后面的语句。

switch 循环语句流程图如图 5-1 所示。

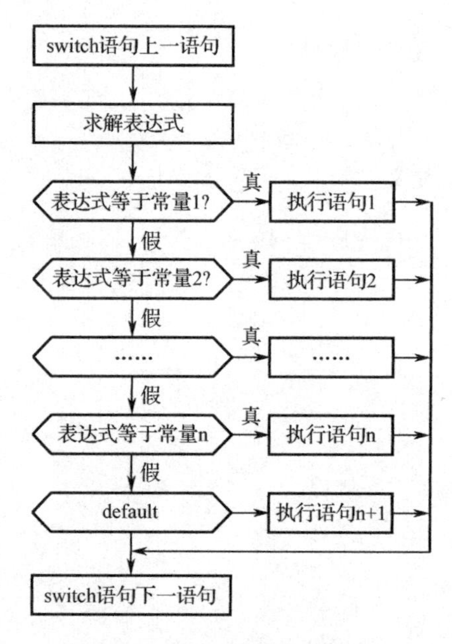

图 5-1　switch 循环语句流程图

5.3　实验步骤

首先，将本书配套资料包的“03.例程资料\Material\04.基于枚举的秒值-时间值转换实验”文件夹复制到 CProgramTest 文件夹中，然后，双击运行“D:\CProgramTest\04.基于枚举的秒值-时间值转换实验\Project”文件夹中的 Project.sln 文件，最后，将程序清单 5-1 中的代码输入 App.c 文件中。下面按照顺序对部分语句进行解释。

（1）第 4 至 11 行代码：定义一个名称为 EnumTimeVal 的枚举类型，然后，使用该枚举类型定义 4 个常量，分别为 TIME_VAL_HOUR、TIME_VAL_MIN、TIME_VAL_SEC 和 TIME_VAL_MAX。

（2）第 17 至 35 行代码：基于枚举和 switch…case…语句，计算小时值、分钟值和秒值。

（3）第 52 至 54 行代码：通过调用 CalcTimeVal 函数计算小时值、分钟值和秒值，具体计算哪个值通过参数传递不同枚举常量来判断。

程序清单 5-1

```
1.   #include <stdio.h>
2.   #include <stdlib.h>
3.
4.   //定义枚举
5.   typedef enum
6.   {
7.     TIME_VAL_HOUR = 0,
```

```
8.    TIME_VAL_MIN,
9.    TIME_VAL_SEC,
10.   TIME_VAL_MAX
11. }EnumTimeVal;
12.
13. //计算小时、分钟和秒值函数声明
14. static  short CalcTimeVal(int tick, unsigned char type);
15.
16.
17. //计算小时、分钟和秒值函数实现
18. static  short CalcTimeVal(int tick, unsigned char type)
19. {
20.   short timeVal;
21.
22.   switch(type)
23.   {
24.   case TIME_VAL_HOUR:
25.     timeVal = tick / 3600;
26.     break;
27.   case TIME_VAL_MIN:
28.     timeVal = (tick % 3600) / 60;
29.     break;
30.   case TIME_VAL_SEC:
31.     timeVal = (tick % 3600) % 60;
32.     break;
33.   default:
34.     break;
35.   }
36.
37.   return(timeVal);
38. }
39.
40. //主函数
41. void main(void)
42. {
43.   int tick = 0;
44.
45.   short hour;
46.   short min;
47.   short sec;
48.
49.   printf("Please input a tick between 0~86399\n");
50.   scanf_s("%d", &tick);
51.
52.   hour = CalcTimeVal(tick, TIME_VAL_HOUR);
53.   min  = CalcTimeVal(tick, TIME_VAL_MIN);
54.   sec  = CalcTimeVal(tick, TIME_VAL_SEC);
55.
56.   printf("Current time: %02d-%02d-%02d\n", hour, min, sec);
57.
58.   system("pause");
59. }
```

最后，按 F5 键编译并运行程序，在弹出的控制台窗口中，输入 80000 后按回车键，可以看到运行结果，即输出“Current time: 22-13-20”，说明实验成功。

本章任务

任务 1：2020 年有 366 天，将 2020 年 1 月 1 日作为计数起点，即计数 1，2020 年 12 月 31 日作为计数终点，即计数 366。计数 1 代表“2020 年 1 月 1 日-星期三”，计数 10 代表“2020 年 1 月 10 日-星期五”。参考本章实验，通过键盘输入一个 1～366 之间的值，包括 1 和 366，基于枚举，将其转换为年、月、日、星期，并输出到控制台窗口。

任务 2：参考本章实验，通过键盘输入 2 个数，判断大小，并输出结果到控制台窗口。

任务 3：参考本章实验，通过键盘输入 10 个数，判断大小，并输出结果到控制台窗口。

本章习题

1．如何使用 typedef 声明新类型名？

2．枚举类型的声明和使用都有哪几种方法？

第6章 基于指针的秒值-时间值转换

指针在C语言中扮演着非常重要的角色，不过它也是一把双刃剑，正确应用可以使程序简洁、紧凑、高效，错误应用则可以让应用程序甚至系统崩溃。本章的实验是基于指针实现秒值-时间值转换，通过该实验，熟悉指针的使用方法。

6.1 实验内容

通过键盘输入一个0～86399之间的值，包括0和86399，使用CalcTimeVal函数计算时间值（包括小时值、分钟值和秒值），函数的输入为tick，输出为指针pTimeVal，即pTimeVal指向时间值，返回值为是否计算成功标志，在主函数中通过调用CalcTimeVal实现秒值-时间值转换，并将结果输出到控制台窗口。

6.2 实验原理

6.2.1 大端模式和小端模式

端模式（Endian）一词出自Jonathan Swift的《格列佛游记》，书中根据敲开鸡蛋的方式不同而将所有人分为两类：从大头端敲开鸡蛋的人被归为Big Endian，从小头端敲开鸡蛋的人被归为Little Endian。小人国的内战就源于吃鸡蛋时究竟是从大头（Big-Endian）敲开，还是从小头（Little-Endian）敲开。在计算机业界，Endian表示数据在存储器中的存放顺序，Big Endian和Little Endian也几乎引发一场战争。下面举例说明在计算机中大、小端模式的区别。

如果有一个32位的int类型的变量为0x12345678，该变量采用大端或小端模式在内存中的存储方式如表6-1所示。

表6-1 大端模式和小端模式在内存中的存储方式

地址偏移	大端模式	小端模式
0x00	12（MSB）	78（LSB）
0x01	34	56
0x02	56	34
0x03	78（LSB）	12（MSB）

如果有一个16位的short类型的变量为0x1234，该变量采用大端或小端模式在内存中的存储方式如表6-2所示。

表6-2 大端模式和小端模式在内存中的存储方式

地址偏移	大端模式	小端模式
0x00	12（MSB）	34（LSB）
0x01	34（LSB）	12（MSB）

由表6-2可知，采用大、小端模式存放数据的主要区别在于存放的字节顺序，大端模式将高位存放在低地址，小端模式将高位存放在高地址。采用大端模式存放数据符合人类的正常思维，采用小端模式存放数据有利于计算机处理。到目前为止，采用大端模式或小端模式进行数据存放，孰优孰劣仍没有定论。

有的处理器系统采用小端模式存放数据，如 Intel 处理器和 ARM 处理器。有的处理器系统采用大端模式存放数据，如 IBM 半导体和 Freescale 的 PowerPC 处理器。不仅对于处理器，一些外设的设计中也存在着使用大端或小端模式进行数据存放的选择。

因此在一个计算机系统中，有可能存在大端模式和小端模式同时存在的现象。这一现象为系统的软、硬件设计带来了不小的麻烦，这就要求系统设计工程师必须深入理解大端和小端模式的差别。大端与小端模式的差别体现在一个处理器的寄存器、指令集和系统总线等各个方面。

6.2.2　常用的三种数据类型的存储方式

计算机的数据类型有很多种，这里只介绍本书中最常用的 3 种数据类型，分别是 unsigned char、short 和 int，这三种数据类型的长度分别为 1 字节、2 字节和 4 字节。

在计算机中，代码、常量、变量和函数等都存储在存储器中，存储器的每个地址存储 1 字节数据。前文提到 Intel 处理器采用的是小端模式，即变量的高位存放在存储器的高地址，本书的所有例程学习都是基于 Intel 处理器的。

为了说明变量的存储方式，下面以程序清单 6-1 中的代码进行介绍。

程序清单 6-1

```
1.  #include <stdlib.h>
2.
3.  void  main(void)
4.  {
5.    int tick = 0;
6.    short arrTimeVal[3];
7.    unsigned char calcRightFlag = 0;
8.
9.    tick = 0x12345678;
10.   arrTimeVal[0] = 0x1122;
11.   arrTimeVal[1] = 0x3344;
12.   arrTimeVal[2] = 0x5566;
13.   calcRightFlag = 0x99;
14.   system("pause");
15. }
```

计算机执行完上述代码后，变量在存储器（Memory）中的存储情况如图 6-1 所示，0x006FFDXX 是存储器的地址，地址右边是存储的变量值，每个地址存储 1 字节数据。注意，这里的地址和数据均为十六进制数。从图 6-1 中可以推导出，0x006FFD43 存储的数据为 0x99，0x006FFD44～0x006FFD4B 存储的数据均为 0xCC（表示存储器初始值），0x006FFD4C 存储的数据为 0x22，0x006FFD4D 存储的数据为 0x11，0x006FFD4E 存储的数据为 0x44，0x006FFD4F 存储的数据为 0x33，0x006FFD50 存储的数据为 0x66，0x006FFD51 存储的数据为 0x55，0x006FFD52～0x006FFD5B 存储的数据为 0xCC，0x006FFD5C 存储的数据为 0x78，0x006FFD5D 存储的数据为 0x56，0x006FFD5E 存储的数据为 0x34，0x006FFD5F 存储的数据为 0x12。

注意，每次在同一台计算机上运行该程序，或在不同的计算机上运行该程序，地址都有可能发生变化。

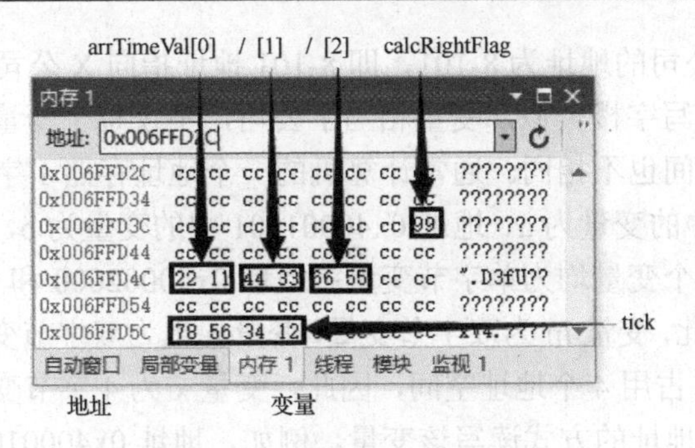

图 6-1　变量对应的存储器

为了更清晰地说明上述代码执行后变量在存储器中的存储方式，将图 6-1 中的地址和变量表示成如图 6-2 所示，左边地址按照从小到大的顺序排列，地址的右边是变量值。从图中可以看出：①在进行本实验的处理器中，存储器按照地址由大到小的顺序存储数据，即优先定义的变量地址大于后定义的变量地址；②数据存储采用小端模式；③存储器未使用的变量区的存储值均为 0xCC。

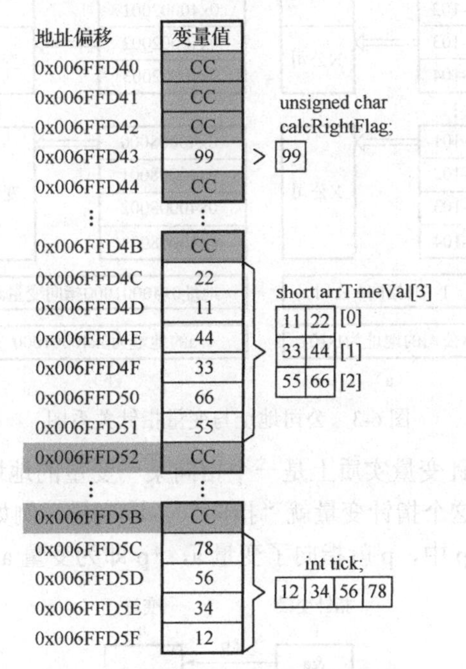

图 6-2　变量在存储器中的存储方式

6.2.3　什么是指针

为了便于介绍指针，借用写字楼的例子。假如一栋写字楼有 8 层，如图 6-3（a）所示，每层都有 4 间房，一楼的房间号依次为 1-101、1-102、1-103 和 1-104，其中，1-101 是 A 公司，1-102 是 B 公司……二楼的公司规模稍微大一些，每个公司占 2 间房，前台（或公司入口）设在其中一间，其中 M 公司的入口设在 2-101，N 公司的入口设在 2-103；位于八层的公司规模更大，占 4 间房，X 公司的入口设在 8-101。当客户来访或快递寄件时，仅仅知道公司名是不可能访问到这些公司的，还需要知道这些公司在该栋楼的具体房间号（也称为地址），例如，A 公司的地址为 1-101，即 1-101 地址指向 A 公司；M 公司的地址为 2-101，即 2-101

地址指向 M 公司；X 公司的地址为 8-101，即 8-101 地址指向 X 公司，等等。

计算机的存储器与写字楼类似，变量相当于公司，不仅每个变量有对应的地址，而且每个变量占用的存储器空间也不相同。通常计算机的一个地址存储 1 字节数据，在图 6-3（b）中，地址 0x40001000 中的变量为 a、地址 0x40001001 中的变量为 b，这两个变量均占用 1 个地址空间，因此，这两个变量均为单字节变量；地址 0x40002000 和 0x40002001 中的变量 m 占用 2 个地址空间，因此，变量 m 为双字节变量，变量 n 也为双字节变量；地址 0x40008000～0x40008003 中的变量 x 占用 4 个地址空间，因此，变量 x 为 4 字节变量。除了通过变量名读写该变量，还可以通过地址的方式读写该变量，例如，地址 0x40001000 存储的变量为 a，地址 0x40001000 指向变量 a。

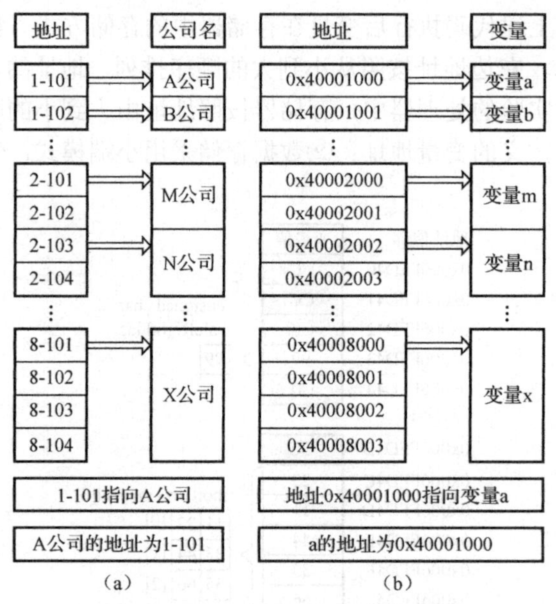

图 6-3　公司地址与变量指针关系图

在 C 语言中，因为指针变量实质上是一个指向某一变量的地址，所以将一个变量的地址值赋给这个指针变量后，这个指针变量就“指向”了该变量。例如，变量 a 的地址为&a，将这个地址存放到指针变量 p 中，p 就指向了变量 a，*p 即为变量 a，如图 6-4 所示。

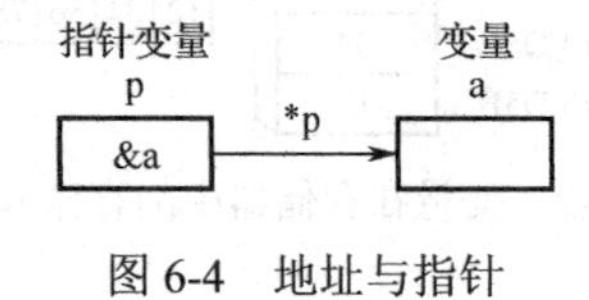

图 6-4　地址与指针

6.2.4　指针变量的定义和使用

定义指针变量的一般形式如下：

```
类型名* 指针变量名;
```

其中，“*”表示定义的是指针变量，类型名为该指针变量所指向的变量的数据类型。

例如：

```
int* pHour;
```

在以下的指针变量使用方法举例中，pHour 为指针变量，该指针变量指向一个整型的变量 hour：

```
int hour;
int* pHour;
pHour = &hour;
```

在以下的指针变量使用方法举例中，指针变量的定义和初始化是一条语句：

```
int hour;
int* pHour = &hour;
```

6.2.5　指针变量注意事项

指针用好了可以编写出优秀的程序，用不好就可能出现 Bug，甚至导致程序崩溃。

使用指针时要注意两点：①指针一定要定义类型，因为指针不仅可以指向单字节变量，还可以指向双字节变量及其他类型的变量，若指针未定义类型则无法使用；②指针在使用前一定要进行初始化，因为未初始化的指针就是野指针，会导致不可预知的后果。如果某一指针指向了内存中比较重要的地方，对该指针进行操作可能会导致系统异常，如系统提示指针指向了一个不可用的地址。因此，指针变量在使用前一定要初始化。

6.2.6　指针与数组

数组名即为数组的地址，也是数组的首地址。例如：

```
unsigned char arr[4] = {0x11, 0x22, 0x33, 0x44};
```

该数组在存储器中的存储方式如图 6-5 所示，arr 即为数组的地址，即 arr = 0x0018FF38，arr[0]的地址也为 0x0018FF38，即&arr[0] = 0x0018FF38。

既然 arr 为数组的地址，而指针就是地址，那么，数组名即为数组的指针。

地址	变量值	unsigned char arr[4]
0x0018FF38	11	[0]
0x0018FF39	22	[1]
0x0018FF3A	33	[2]
0x0018FF3B	44	[3]

图 6-5　数组在存储器中的存储方式

数组和指针的对应关系如表 6-3 所示，左右等效。

表 6-3　数组和指针的对应关系

数组操作	指针操作
&arr[0]	arr
arr[0]	*arr
arr[0]	*(arr+0)
arr[1]	*(arr+1)
arr[2]	*(arr+2)
arr[3]	*(arr+3)

6.2.7　用 if 语句实现选择结构

if 语句是最简单的选择流程语句，C 语言中的选择结构主要是由 if 语句实现的，最常用的有三种形式。

（1）形式一

```
if(表达式)
{
    语句 1;
}
```

（2）形式二

```
if(表达式)
{
    语句 1;
}
else
{
    语句 2;
}
```

（3）形式三

```
if(表达式 1)
{
    语句 1;
}
else if(表达式 2)
{
    语句 2;
}
else if(表达式 3)
{
    语句 3;
}
…
else if(表达式 n)
{
    语句 n;
}
else
{
    语句 n+1;
}
```

在 if 语句中要先判断给定的表达式：若表达式成立，即逻辑为真，则执行对应花括号中的语句；若表达式不成立，即逻辑为假，则跳过花括号中的语句继续判断下一个表达式；若所有表达式都不成立，则执行 else 后面的语句。

“语句 1”“语句 2”…“语句 n+1”可以是一个单独的语句，也可以是一个包含多个语句的复合语句。

通常在编写代码时，容易将 if(a == 1)误写为 if(a = 1)，这样就会引入 Bug。这是由于 a = 1

是赋值语句，因此 a = 1 恒成立，例如，执行完语句“b = (a = 1);”后的结果就是 b = 1。这样，if(a = 1)即为 if(1)，if(a = 1)条件下的代码将会无条件执行，与判断语句 if(a == 1)的执行结果相违背，更重要的是一般的编译器不会报 error 或 warnning。

为了避免出现上述 Bug，本书建议将 if(a == 1)写为 if(1 == a)，此时若误将 if(1 == a)写为 if(1 = a)，编译器就会报 error。

注意，为了编写出高效且优质的代码，本书会提出很多建议来规避一些常见错误。

6.2.8　逻辑运算符

逻辑运算符有三种：与（AND）、或（OR）、非（NOT），在 C 语言中分别用符号“&&”“||”和“!”来表示，如表 6-4 所示。

表 6-4　逻辑运算符及其含义

逻辑运算符	含　义	示　例	说　明
&&	逻辑与	m&&n	若 m 和 n 都为真，则结果为真，否则为假
||	逻辑或	m||n	若 m 和 n 都为假，则结果为假，否则为真
!	逻辑非	!m	若 m 为真，则结果为假；若 m 为假，则结果为真

6.2.9　程序调试

调试，就是在程序运行过程中的某一阶段观测程序运行的状态，而程序通常是连续运行的，所以必须使程序在某一点停下来。首先，设置断点；其次，运行程序；最后，当程序在断点处停下来时，观察程序运行的状态。

下面将通过实验来说明调试过程。第一步，在“calcRightFlag = CalcTimeVal(tick, arrTimeVal);”代码所在行设置断点，具体做法是单击此行代码，按 F9 键，此时该行代码左侧会出现红色圆点，如图 6-6 所示，表示在此行设置了断点。注意，如果要取消断点，单击此行代码，再次按 F9 键，红点消失表示断点取消。

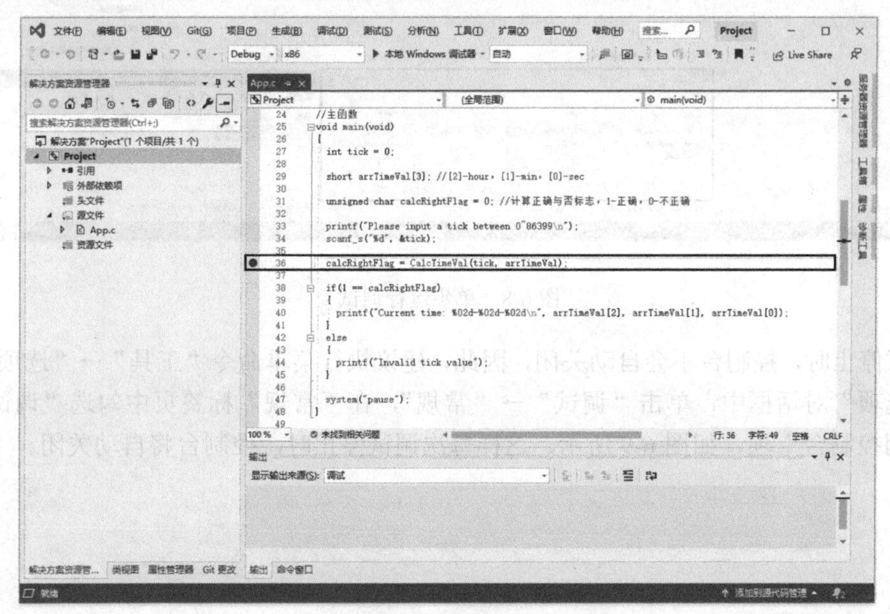

图 6-6　设置断点

第二步，按 F5 键运行程序。由于程序中有 scanf_s 语句，需要用户输入 tick 值，而断点正好在 scanf_s 语句之后，因此，运行程序后系统会弹出如图 6-7 所示的窗口。在窗口中输入一个数值后，程序才能运行到断点处，这里输入 80000 后按回车键，最后将图 6-7 所示的窗口最小化。

图 6-7　输入 tick 值

随后系统会弹出如图 6-8 所示的界面，黄色箭头指示当前程序运行到的地方。第三步，按 F11 键，单步运行程序。如果黄色箭头指示的是函数调用代码行，按 F11 键，程序会跳转至函数实现区；否则会单步向下执行。在局部变量区可以查看变量取值，例如，calcRightFlag 等于 0，tick 等于 80000，如图 6-8 所示。如果要停止调试，按 Shift+F5 组合键即可。

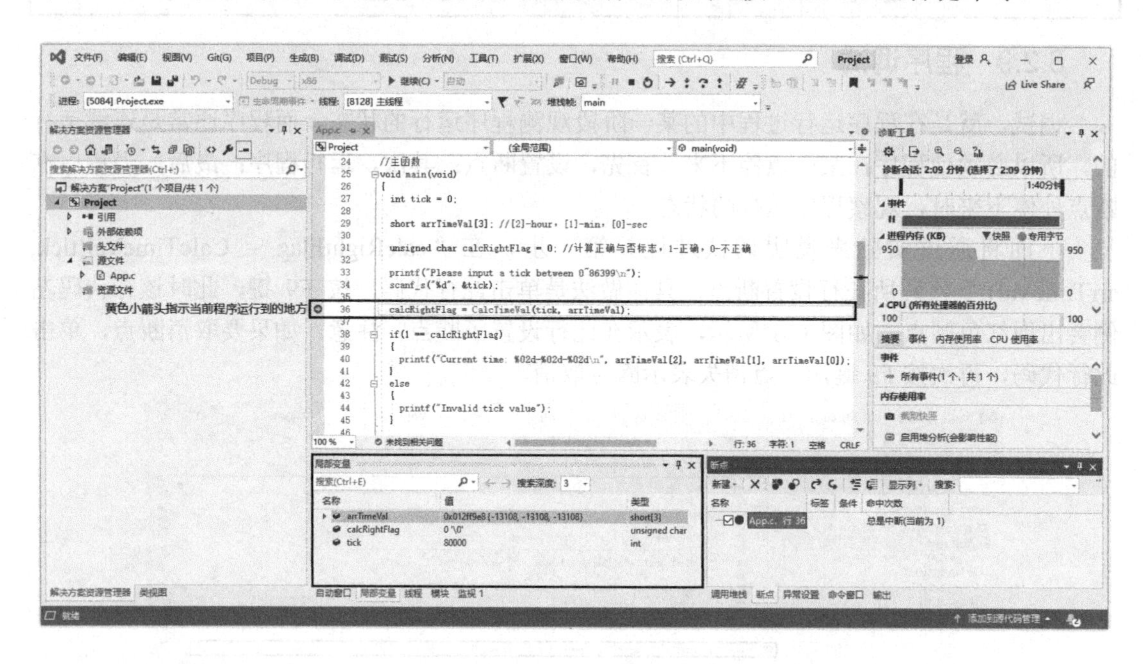

图 6-8　单步运行调试

调试停止时，控制台不会自动关闭，因此，建议执行菜单命令“工具”→“选项”，在弹出的“选项”对话框中，单击“调试”→“常规”，在“常规”标签页中勾选“调试停止时自动关闭控制台”项，如图 6-9 所示，这样每次调试停止时，控制台将自动关闭。

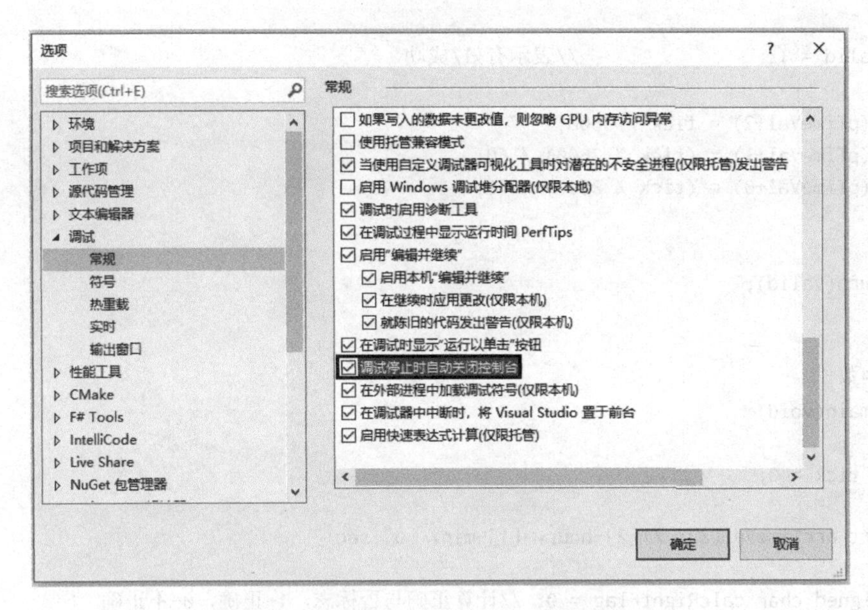

图 6-9　设置调试停止时自动关闭控制台

6.3　实验步骤

首先，将本书配套资料包的"03.例程资料\Material\05.基于指针的秒值-时间值转换实验"文件夹复制到 CProgramTest 文件夹中，然后，双击运行"D:\CProgramTest\05.基于指针的秒值-时间值转换实验\Project"文件夹中的 Project.sln 文件，最后，将程序清单 6-1 中的代码输入 App.c 文件中。下面按照顺序对部分语句进行解释。

（1）第 5 至 22 行代码：声明和定义计算小时、分钟和秒值的 CalcTimeVal 函数，该函数有 2 个参数，分别为 tick 和 pTimeVal，其中，pTimeVal 是一个 short 型的指针变量。

（2）第 29 行代码：声明一个 short 型数组，数组名为 arrTimeVal。

（3）第 36 行代码：将 arrTimeVal 作为参数传入 CalcTimeVal 函数，arrTimeVal 既是数组名，也是数组第一个元素的地址，因此，第一个元素既可以通过数组方式表示，即 arrTimeVal[0]，也可以用指针方式表示，即*(arrTimeVal+0)，表示 arrTimeVal[0]等效于*(arrTimeVal+0)，arrTimeVal[1]等效于*(arrTimeVal+1)，arrTimeVal[2]等效于*(arrTimeVal+2)，因此，第 16 至 18 行代码左边的值也可以用数组方式表示。

程序清单 6-1

```
1.   #include <stdio.h>
2.   #include <stdlib.h>
3.
4.   //计算小时、分钟和秒值函数声明
5.   static  unsigned char CalcTimeVal(int tick, short* pTimeVal);
6.
7.   //计算小时、分钟和秒值函数实现，返回值表示是否计算成功，0-失败，1-成功
8.   static  unsigned char CalcTimeVal(int tick, short* pTimeVal)
9.   {
10.    unsigned char valid = 0;     //初始值为 0，即表示无效/失败
11.
12.    if(tick >= 0 &&  tick <= 86399)
```

```
13.    {
14.      valid = 1;                    //表示有效/成功
15.
16.      *(pTimeVal+2) = tick / 3600;
17.      *(pTimeVal+1) = (tick % 3600) / 60;
18.      *(pTimeVal+0) = (tick % 3600) % 60;
19.    }
20.
21.    return(valid);
22.  }
23.
24.  //主函数
25.  void main(void)
26.  {
27.    int tick = 0;
28.
29.    short arrTimeVal[3]; //[2]-hour, [1]-min, [0]-sec
30.
31.    unsigned char calcRightFlag = 0; //计算正确与否标志，1-正确，0-不正确
32.
33.    printf("Please input a tick between 0~86399\n");
34.    scanf_s("%d", &tick);
35.
36.    calcRightFlag = CalcTimeVal(tick, arrTimeVal);
37.
38.    if(1 == calcRightFlag)
39.    {
40.      printf("Current time: %02d-%02d-%02d\n", arrTimeVal[2], arrTimeVal[1], arrTimeVal[0]);
41.    }
42.    else
43.    {
44.      printf("Invalid tick value");
45.    }
46.
47.    system("pause");
48.  }
```

最后，按 F5 键编译并运行程序，在弹出的控制台窗口中，输入 80000 后按回车键，可以看到运行结果，即输出“Current time: 22-13-20”，说明实验成功。

本 章 任 务

任务 1：2020 年有 366 天，将 2020 年 1 月 1 日作为计数起点，即计数 1，2020 年 12 月 31 日作为计数终点，即计数 366。计数 1 代表“2020 年 1 月 1 日-星期三”，计数 10 代表“2020 年 1 月 10 日-星期五”。参考本章实验，通过键盘输入一个 1～366 之间的值，包括 1 和 366，基于指针，将其转换为年、月、日、星期，并输出到控制台窗口。

任务 2：参考本章实验，通过键盘输入 2 个数，判断大小，并输出结果到控制台窗口。

任务 3：参考本章实验，通过键盘输入 10 个数，判断大小，并输出结果到控制台窗口。

本 章 习 题

1．什么是大端模式和小端模式？

2．什么是指针？如何定义指针变量和使用指针？

3．简述指针与数组之间的关系。

4．简述程序调试的过程。

第 7 章　基于结构体的秒值-时间值转换

结构体是由一批数据组合而成的数据类型，组成结构型数据的每个数据称为结构型数据的“成员”。本章的实验基于结构体实现秒值-时间值转换，通过该实验，熟悉结构体的声明和使用方法。

7.1　实验内容

通过键盘输入一个 0～86399 之间的值，包括 0 和 86399，使用 CalcTimeVal 函数计算时间值，函数的输入为 tick，函数的返回值为一个结构体变量，结构体的成员包括小时值、分钟值和秒值，在主函数中通过调用 CalcTimeVal 实现秒值-时间值转换，并将结果输出到控制台窗口。

7.2　实验原理

7.2.1　为什么要使用结构体类型

有些数据之间有内在联系，成组出现，如一个学生的学号（num）、姓名（name）、性别（sex）和年龄（age）等信息，都属于同一个学生。如果将 num、name、sex、age 简单地定义为相互独立的变量，则无法反映其中的内在联系；如果定义一个变量，该变量中包含学生的学号、姓名、性别和年龄等项，这样就能显示出这些变量之间的关联。

7.2.2　结构体类型的声明和使用方法

结构体类型的声明和使用方法有三种。

（1）方法一

```
struct StructStudent{
int num;
char name[20];
char sex;
int age
}; //声明结构体
StructStudent student1; //定义变量
//使用
student1.num = 1234;
student2.name = “zhangsan”;
student3.sex = ‘M’;
student4.age = 20;
```

（2）方法二

```
struct {
int num;
char name[20];
char sex;
```

```
int age
} student1; //声明、定义
//使用
student1.num = 1234;
student2.name = “zhangsan”;
student3.sex = ‘M’;
student4.age = 20;
```

（3）方法三

```
//声明结构体类型
typedef struct {
int num;
char name[20];
char sex;
int age
} StructStudent ;
StructStudent student1; //定义
//使用
student1.num = 1234;
student2.name = “zhangsan”;
student3.sex = ‘M’;
student4.age = 20;
```

与枚举类似，本书建议采用第三种声明和使用方法。

7.3 实验步骤

首先，将本书配套资料包的“03.例程资料\Material\06.基于结构体的秒值-时间值转换实验”文件夹复制到 CProgramTest 文件夹中，然后，双击运行“D:\CProgramTest\06.基于结构体的秒值-时间值转换实验\Project”文件夹中的 Project.sln 文件，最后，将程序清单 7-1 中的代码输入 App.c 文件中。下面按照顺序对部分语句进行解释。

（1）第 4 至 10 行代码：声明一个名称为 StructTimeVal 的结构体类型，该结构体包含 3 个成员变量，分别为 hour、min 和 sec。

（2）第 12 至 25 行代码：声明和定义计算小时、分钟和秒值的 CalcTimeVal 函数，返回值的类型为结构体。

（3）第 37 行代码：通过调用 CalcTimeVal 函数计算小时值、分钟值和秒值，结果保存在结构体变量 tv 中。

程序清单 7-1

```
1.   #include <stdio.h>
2.   #include <stdlib.h>
3.
4.   //声明一个时间值结构体，包括三个成员变量，分别是 hour，min 和 sec
5.   typedef struct
6.   {
7.     short hour;
8.     short min;
9.     short sec;
10.  }StructTimeVal;
11.
```

```
12. //计算小时、分钟和秒值函数声明
13. static  StructTimeVal CalcTimeVal(int tick);
14.
15. //计算小时、分钟和秒值，返回值为一个结构体类型的变量
16. static  StructTimeVal CalcTimeVal(int tick)
17. {
18.   StructTimeVal tv;
19.
20.   tv.hour = tick / 3600;
21.   tv.min  = (tick % 3600) / 60;
22.   tv.sec  = (tick % 3600) % 60;
23.
24.   return(tv);
25. }
26.
27. //主函数
28. void main(void)
29. {
30.   int tick = 0;
31.
32.   StructTimeVal tv;
33.
34.   printf("Please input a tick between 0~86399\n");
35.   scanf_s("%d", &tick);
36.
37.   tv = CalcTimeVal(tick);
38.
39.   printf("Current time: %02d-%02d-%02d\n", tv.hour, tv.min, tv.sec);
40.
41.   system("pause");
42. }
```

最后，按 F5 键编译并运行程序，在弹出的控制台窗口中，输入 80000 后按回车键，可以看到运行结果，即输出“Current time: 22-13-20”，说明实验成功。

本 章 任 务

任务 1：2020 年有 366 天，将 2020 年 1 月 1 日作为计数起点，即计数 1，2020 年 12 月 31 日作为计数终点，即计数 366。计数 1 代表“2020 年 1 月 1 日-星期三”，计数 10 代表“2020 年 1 月 10 日-星期五”。参考本章实验，通过键盘输入一个 1～366 之间的值，包括 1 和 366，基于结构体，将其转换为年、月、日、星期，并输出到控制台窗口。

任务 2：参考本章实验，通过键盘输入 2 个数，判断大小，并输出结果到控制台窗口。

任务 3：参考本章实验，通过键盘输入 10 个数，判断大小，并输出结果到控制台窗口。

本 章 习 题

1．为什么要使用结构体类型？

2．简述结构体类型的声明和使用方法。

第 8 章　基于结构体指针的秒值-时间值转换

第 7 章通过“结构体变量名.成员名”的方式引用结构体变量中的成员，除了这种方法，还可以使用结构体指针。本章的实验基于结构体指针实现秒值-时间值转换，通过该实验，熟悉结构体指针的使用方法。

8.1　实验内容

通过键盘输入一个 0～86399 之间的值，包括 0 和 86399，使用 CalcTimeVal 函数计算时间值，函数的输入为 tick，输出为指针 pTV，pTV 指向一个结构体变量，结构体的成员包括小时值、分钟值和秒值，返回值为是否计算成功标志，在主函数中通过调用 CalcTimeVal 函数实现秒值-时间值转换，并将结果输出到控制台窗口。

8.2　实验原理

8.2.1　结构体指针

结构体指针就是指向结构体变量的指针，即为结构体变量的起始地址。首先，声明一个结构体，如下所示：

```
typedef struct
{
  short hour;
  short min;
  short sec;
}StructTimeVal;
```

再用 StructTimeVal 定义一个结构体类型的指针变量，如下所示：

```
StructTimeVal* pTV;
```

前文提到，指针在使用前一定要初始化，结构体指针也一样。例如，有一个结构体变量如下所示：

```
StructTimeVal timeVal;
```

使得 pTV 指向上面的结构体变量，如下所示：

```
pTV = &timeVal;
```

结构体用“结构体变量名.成员名”的方式引用结构体变量中的成员，结构体指针用“结构体指针名->成员名”的方式引用结构体变量中的成员，如 pTV->hour、PTV->min、pTV->sec。

8.2.2　指针加 *M* 操作

指针即为地址，指向某一变量。如果指针 pA 指向的变量的长度为 *N* 字节，那么指针 pA 加 1 操作即为 pA 的地址+*N*，pA 加 *M* 操作即为 pA 的地址+*N*×*M*。

如图 8-1 所示，当指针指向 unsigned char 型变量（单字节变量）时，如 unsigned char* pA，pA+1 操作即为 pA 的地址+1，pA+*M* 操作即为 pA 的地址+*M*；当指针指向 short 型变量（双字节变量）时，如 short* pA，pA+1 操作即为 pA 的地址+2，pA+*M* 操作即为 pA 的地址+2×*M*；当指针指向 int 型变量（4 字节变量）时，如 int* pA，pA+1 操作即为 pA 的地址+4，pA+*M* 操作即为 pA 的地址+4×*M*。

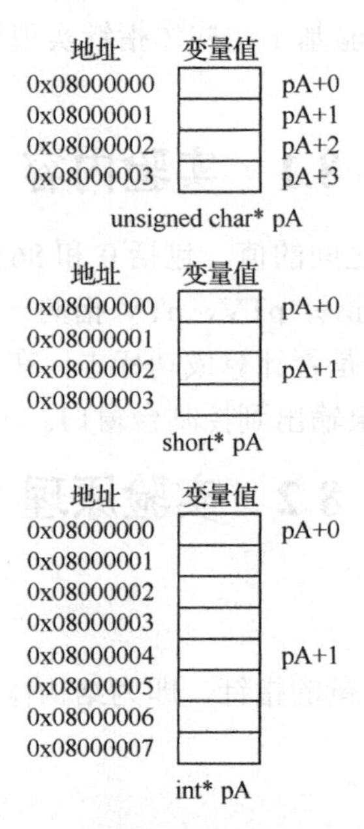

图 8-1　变量指针变量加 1 操作

如图 8-2 所示的结构体 StructTimeVal，当指针指向 StructTimeVal 结构体变量（6 字节变量）时，如 StructTimeVal* pA，pA+1 操作即为 pA 的地址+6，pA+*M* 操作即为 pA 的地址+6×*M*。

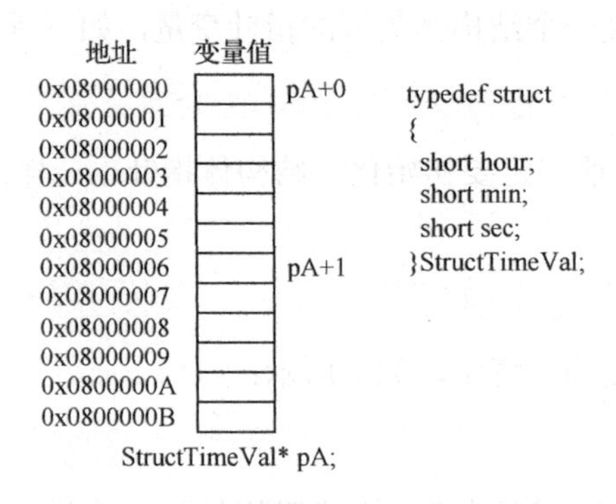

图 8-2　结构体指针变量加 1 操作

8.3　实验步骤

首先，将本书配套资料包的“03.例程资料\Material\07.基于结构体指针的秒值-时间值转换实验”文件夹复制到 CProgramTest 文件夹中，然后，双击运行“D:\CProgramTest\07.基于结构体指针的秒值-时间值转换实验\Project”文件夹中的 Project.sln 文件，最后，将程序清单 8-1 中的代码输入 App.c 文件中。下面按照顺序对部分语句进行解释。

（1）第 13 至 29 行代码：声明和定义计算小时、分钟和秒值的 CalcTimeVal 函数，该函数的 pTV 参数是一个结构体类型的指针变量，因此，引用成员变量使用“->”，即 pTV->hour、pTV->min、pTV->sec 分别表示小时值、分钟值和秒值变量。

（2）第 43 行代码：将结构体变量 tv 的地址，即 &tv 作为参数传入 CalcTimeVal 函数，这样，当 CalcTimeVal 函数执行结束后，计算的小时值、分钟值和秒值将被保存在 tv 中。

程序清单 8-1

```
1.  #include <stdio.h>
2.  #include <stdlib.h>
3.
4.  //定义一个时间值结构体，包括三个成员变量，分别是 hour，min 和 sec
5.  typedef struct
6.  {
7.    short hour;
8.    short min;
9.    short sec;
10. }StructTimeVal;
11.
12. //计算小时、分钟和秒值函数声明
13. static  unsigned char CalcTimeVal(int tick, StructTimeVal* pTV);
14.
15. //计算小时、分钟和秒值函数实现，返回值表示是否计算成功，0-失败，1-成功
16. static  unsigned char CalcTimeVal(int tick, StructTimeVal* pTV)
17. {
18.   unsigned char valid = 0;
19.
20.   if(tick >= 0 &&  tick <= 86399)
21.   {
22.     valid = 1;
23.     pTV->hour = tick / 3600;
24.     pTV->min  = (tick % 3600) / 60;
25.     pTV->sec  = (tick % 3600) % 60;
26.   }
27.
28.   return(valid);
29. }
30.
31. //主函数
32. void main(void)
33. {
34.   int tick = 0;
35.
36.   unsigned char valid;
37.
38.   StructTimeVal tv;
39.
40.   printf("Please input a tick between 0~86399\n");
41.   scanf_s("%d", &tick);
42.
43.   valid = CalcTimeVal(tick, &tv);
```

```
44.
45.    if(valid > 0)
46.    {
47.      printf("Current time: %02d-%02d-%02d\n", tv.hour, tv.min, tv.sec);
48.    }
49.    else
50.    {
51.      printf("Invalid tick value!!!\n");
52.    }
53.
54.    system("pause");
55. }
```

最后，按 F5 键编译并运行程序，在弹出的控制台窗口中，输入 80000 后按回车键，可以看到运行结果，即输出“Current time: 22-13-20”，说明实验成功。

本 章 任 务

任务 1：2020 年有 366 天，将 2020 年 1 月 1 日作为计数起点，即计数 1，2020 年 12 月 31 日作为计数终点，即计数 366。计数 1 代表“2020 年 1 月 1 日-星期三”，计数 10 代表“2020 年 1 月 10 日-星期五”。参考本章实验，通过键盘输入一个 1～366 之间的值，包括 1 和 366，基于结构体指针，将其转换为年、月、日、星期，并输出到控制台窗口。

任务 2：参考本章实验，通过键盘输入 2 个数，判断大小，并输出结果到控制台窗口。

任务 3：参考本章实验，通过键盘输入 10 个数，判断大小，并输出结果到控制台窗口。

本 章 习 题

1．什么是结构体指针？

2．简述结构体指针的操作。

3．变量指针加 1 操作和结构体指针加 1 操作的区别是什么？

第9章　基于多文件的秒值-时间值转换

做一些小实验往往只需要一个包含 main 函数的 C 文件即可。然而对于一些复杂的项目，就需要按照功能将其拆分为多个文件，每个文件扮演不同的角色，最终由包含 main 函数的 C 文件直接或间接调用其他文件中的函数，从而实现某种功能。本章的实验基于多文件实现秒值-时间值转换，通过该实验，熟悉多文件的使用方法。

9.1　实验内容

通过键盘输入一个 0～86399 之间的值，包括 0 和 86399，使用 CalcHour.c/.h 文件对计算小时值，使用 CalcMin.c/.h 文件对计算分钟值，使用 CalcSec.c/.h 文件对计算秒值，在主函数中通过调用 CalcHour、CalcMin 和 CalcSec 模块中相关的函数实现秒值-时间值转换，并将结果输出到控制台窗口。

9.2　实验原理

9.2.1　项目架构

前面几章的实验都以一个 App.c 文件完成所有的功能，而本章的实验将使用多文件实现秒值-时间值转换，因此，就需要有一个清晰的项目架构。

本章实验的项目架构如图 9-1 所示，App.c 可调用 CalcHour.c/.h、CalcMin.c/.h 和 CalcSec.c/.h 三个文件对，.h 文件为头文件，主要用于声明 API 函数，即被其他文件调用的全局函数，.c 文件用于定义 API 函数，即实现 API 函数的具体功能。

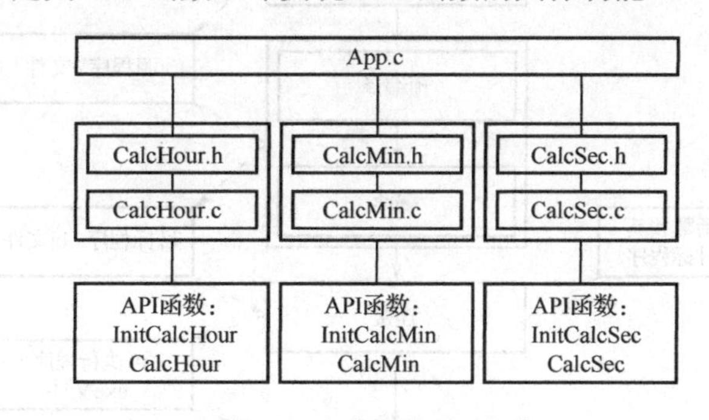

图 9-1　本章实验项目架构

9.2.2　为什么要使用多文件

前面在介绍函数时，提到过模块化设计思想，本章实验将一个工程分成若干文件正是体现了该思想。模块化设计主要有以下几个优点。

（1）便于代码复用。通用性强或功能重复的模块只需写一遍，可被反复调用，如一个加法计算单元或独立按键去抖模块。

（2）便于团队协作。系统工程师只需要从宏观上将一个项目划分为若干小模块，然后明确每个小模块之间的接口，程序员即可独立完成各自负责的模块，最后再由系统工程师将各个模块组合为一个整体。

（3）便于修改和维护。当某个模块出现问题时，只需要检查并完善该模块。另外，当需要升级程序的某一功能时，可以只针对该功能对应的模块重新开发，既省时又省力。

9.2.3　编译过程

在学习程序的编译过程之前，先了解几个基本概念。

（1）编辑器：用于编写代码，如记事本、Word 和 Notepad 等。

（2）编译器：用于检查代码的语法错误并将其编译成二进制形式的目标程序。

（3）连接器：用于将目标程序连接成可执行文件。

（4）集成开发环境（Integrated Development Environment，IDE）：用于程序开发的应用程序，通常包括代码编辑器、编译器、调试器和图形用户界面工具，如 Visual Studio 2019 等。

下面介绍在 Visual Studio 2019 集成开发环境下编译程序的过程，如图 9-2 所示。

（1）用编辑器编辑源程序，输出源程序文件（.c 文件）。

（2）编译源程序，先用预编译器对预处理指令进行编译预处理。例如，对于#include <stdio.h>指令来说，就是将 stdio.h 头文件中的内容读出来，并取代#include <stdio.h>行，由预处理得到的信息与程序的其他部分一起组成一个完整的、可以正式编译的源程序，然后由编译系统编译该程序，最终生成二进制形式的目标程序（.obj 文件）。

（3）用连接器将二进制目标文件、库函数及其他目标程序连接成一个整体，生成一个可供计算机执行的目标程序，称为可执行程序（.exe 文件）。

（4）最后，运行可执行程序，得到运行结果。

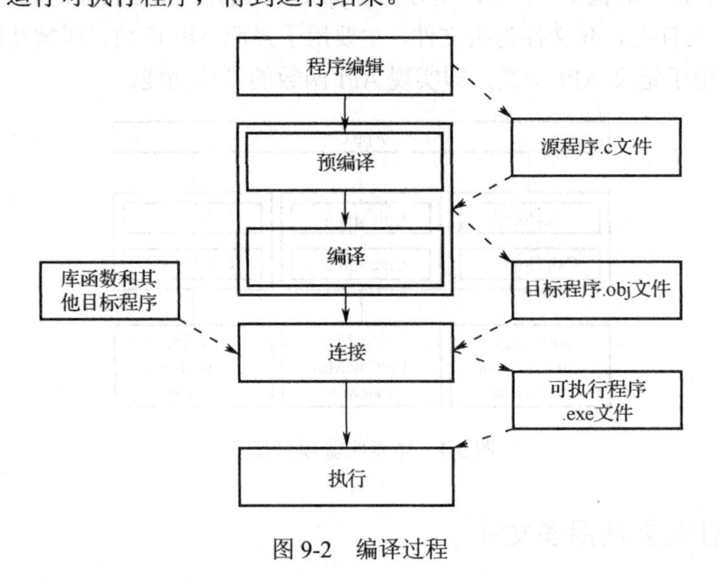

图 9-2　编译过程

9.2.4　防止重编译

在前面提到过预处理，下面介绍重编译的概念及如何防止重编译。如图 9-3 所示，头文件 Add.h 被 File1.c 和 File2.c 调用，经过预处理后，由于#include "Add.h"被头文件 Add.h 的内容“int Add(int a, int b);”替换，因此，在编译时，“int Add(int a, int b);”就会被编译两次，这就是重编译。针对重编译的问题，有些编译器会报错，但有些不会，为了增强程序的健壮性，建议进行防止重编译处理。

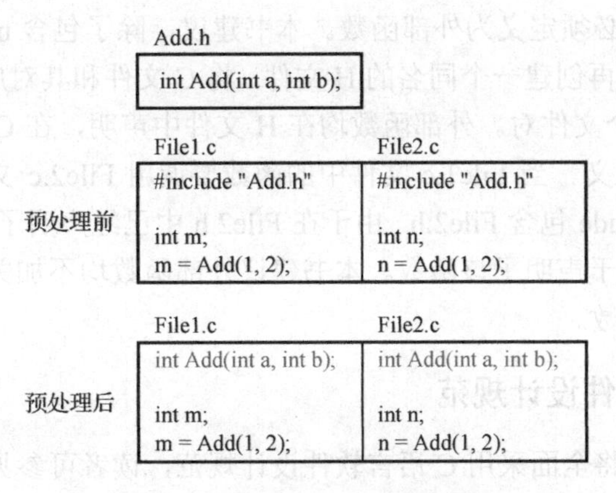

图 9-3　未进行防止重编译处理

防止重编译的具体做法如图 9-4 所示，在 Add.h 文件的代码前后分别添加预处理指令。这样，在编译预处理 File1.c 时，预处理器将#include "Add.h"用“int Add(int a, int b);”替换，同时定义了_ADD_H_，随后编译预处理 File2.c 时，发现_ADD_H_已经被定义，因此，就不再用“int Add(int a, int b);”替换#include "Add.h"，而是直接将#include "Add.h"删除，这样就起到了防止重编译的效果。

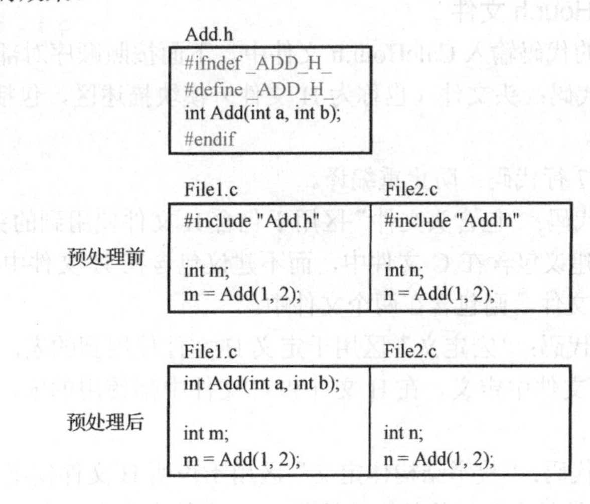

图 9-4　防止重编译处理

本书建议所有的头文件都添加防止重编译代码。

9.2.5　API 函数

与内部函数对应的是外部函数，声明外部函数的一般格式如下：

```
extern 类型名 函数名(形参列表);
```

外部函数可供其他文件调用。C 语言规定，如果在声明函数时省略 extern，则默认为外部函数。在需要调用此函数的其他文件中，需要对此函数作声明并加关键字 extern，表示该函数为其他文件中定义的外部函数。

但考虑到程序设计规范，本书建议内部函数均在 C 文件中声明和定义，且无论是声明还是定义，都不要省略 static。另外，仅在文件内部使用的函数必须定义为内部函数，在其他文件中将会被调用的函数必须定义为外部函数。本书建议，除了包含 main 函数的 App.c 文件，其余每个 C 文件都必须再创建一个同名的 H 文件，将 C 文件和其对应的 H 文件称为文件对，如 File.h 和 File.c 为一个文件对。外部函数均在 H 文件中声明，在 C 文件中定义；而内部函数在 C 文件中声明并定义。当 File1.c 文件中的函数要调用 File2.c 文件中的函数时，必须在 File1.c 文件中通过#include 包含 File2.h。由于在 File2.h 中已经声明了要被调用的函数，因此，File1.c 文件中也就相当于声明了该函数。本书建议外部函数均不加关键字 extern，这样也就将外部函数称为 API 函数。

9.2.6　C 语言软件设计规范

从本章实验开始，将全面采用 C 语言软件设计规范，读者可参见附录 A。

9.3　实验步骤

步骤 1：复制原始项目

首先，将本书配套资料包的“03.例程资料\Material\08.基于多文件的秒值-时间值转换实验”文件夹复制到 CProgramTest 文件夹中，然后，双击运行“D:\CProgramTest\08.基于多文件的秒值-时间值转换实验\Project”文件夹中的 Project.sln 文件。

步骤 2：完善 CalcHour.h 文件

将程序清单 9-1 中的代码输入 CalcHour.h 文件中。下面按照顺序对部分语句进行解释。

（1）第 1 至 15 行代码：头文件（也称为 H 文件）模块描述区，包括模块名称、摘要、当前版本和作者等信息。

（2）第 16、17 和 37 行代码：防止重编译。

（3）第 19 至 22 行代码：“包含头文件”区用于包含 H 文件使用到的头文件，对于只在 C 文件中使用的头文件，建议包含在 C 文件中，而不建议包含在 H 文件中。当然，若在 H 文件和 C 文件中都使用头文件，则包含在两个文件中。

（4）第 23 至 26 行代码：“宏定义”区用于定义 H 文件使用到的宏，对于只在 C 文件中使用的宏，建议只在 C 文件中定义。在 H 文件和 C 文件中都使用的宏，则在两个文件中都定义。

（5）第 27 至 30 行代码：“枚举结构体定义”区用于声明 H 文件使用到的枚举和结构体，对于只在 C 文件中使用的枚举和结构体，建议只在 C 文件中定义。在 H 文件和 C 文件中都使用的枚举和结构体，则在两个文件中都定义。

（6）第 31 至 35 行代码：“API 函数声明”区用于声明所有将被外部文件调用的函数。

程序清单 9-1

```
1.   /*********************************************************************************
2.   * 模块名称：CalcHour.h
3.   * 摘    要：计算小时
4.   * 当前版本：1.0.0
5.   * 作    者：Leyutek(COPYRIGHT 2018 - 2021 Leyutek. All rights reserved.)
6.   * 完成日期：2021 年 03 月 01 日
7.   * 内    容：
```

```
8.  * 注    意:
9.  ********************************************************************************
10. * 取代版本:
11. * 作    者:
12. * 完成日期:
13. * 修改内容:
14. * 修改文件:
15. ********************************************************************************/
16. #ifndef _CALC_HOUR_H_
17. #define _CALC_HOUR_H_
18.
19. /*******************************************************************************
20. *                                包含头文件
21. ********************************************************************************/
22.
23. /*******************************************************************************
24. *                                宏定义
25. ********************************************************************************/
26.
27. /*******************************************************************************
28. *                                枚举结构体定义
29. ********************************************************************************/
30.
31. /*******************************************************************************
32. *                                API 函数声明
33. ********************************************************************************/
34. void  InitCalcHour(void);         //初始化 CalcHour 模块
35. short CalcHour(int tick);         //计算小时值
36.
37. #endif
```

步骤 3：完善 CalcHour.c 文件

将程序清单 9-2 中的代码输入 CalcHour.c 文件中。下面按照顺序对这些语句进行解释。

（1）第 1 至 15 行代码：C 文件模块描述区，包括模块名称、摘要、当前版本和作者等信息，与 H 文件模块描述区类似。

（2）第 17 至 21 行代码："包含头文件"区首先要包含 C 文件对应的 H 文件，即#include "CalcHour.h"，其次，要包含仅在 C 文件中使用，而 H 文件中不使用的头文件。

（3）第 22 至 25 行代码："宏定义"区中定义的宏，应是仅在 C 文件中使用的宏。

（4）第 26 至 29 行代码："枚举结构体定义"区中定义的枚举和结构体，应是仅在 C 文件中使用的枚举和结构体。

（5）第 30 至 33 行代码："内部变量"区用于定义 C 文件中的函数之间所使用的变量，须加 static 关键字，变量名以 s_开头。建议不要在其他文件中直接调用另一个文件中的变量，如果需要调用，则必须使用函数的形式，例如，通过 GetXyz 函数获取其他文件中的变量，通过 SetXyz 函数更改其他文件中的变量。

（6）第 34 至 37 行代码："内部函数声明"区用于声明所有内部函数，对于只会被其他文件中的函数调用的函数，不允许在"内部函数声明"区中声明。

（7）第 38 至 41 行代码："内部函数实现"区用于定义内部函数。

（8）第 42 至 74 行代码："API 函数实现"区用于定义外部函数，虽然 InitCalcHour 函数体为空，但这个函数依然要保留，并且建议在编写程序时，每个模块都预留一个初始化函数。例如，CalcHour 模块要有一个 InitCalcHour 函数，ProcMCUData 模块要有一个 InitProcMCUData 函数。将函数提前设置好（即使为空函数），以便于后期升级。

程序清单 9-2

```
1.  /*********************************************************************************
2.  * 模块名称: CalcHour.c
3.  * 摘    要: 计算小时
4.  * 当前版本: 1.0.0
5.  * 作    者: Leyutek(COPYRIGHT 2018 - 2021 Leyutek. All rights reserved.)
6.  * 完成日期: 2021 年 03 月 01 日
7.  * 内    容:
8.  * 注    意:
9.  **********************************************************************************
10. * 取代版本:
11. * 作    者:
12. * 完成日期:
13. * 修改内容:
14. * 修改文件:
15. *********************************************************************************/
16.
17. /*********************************************************************************
18. *                                  包含头文件
19. *********************************************************************************/
20. #include "CalcHour.h"
21.
22. /*********************************************************************************
23. *                                  宏定义
24. *********************************************************************************/
25.
26. /*********************************************************************************
27. *                                  枚举结构体定义
28. *********************************************************************************/
29.
30. /*********************************************************************************
31. *                                  内部变量
32. *********************************************************************************/
33.
34. /*********************************************************************************
35. *                                  内部函数声明
36. *********************************************************************************/
37.
38. /*********************************************************************************
39. *                                  内部函数实现
40. *********************************************************************************/
41.
42. /*********************************************************************************
43. *                                  API 函数实现
44. *********************************************************************************/
45. /*********************************************************************************
```

```
46.  * 函数名称: InitCalcHour
47.  * 函数功能: 初始化小时计算模块
48.  * 输入参数: void
49.  * 输出参数: void
50.  * 返回值: void
51.  * 创建日期: 2021 年 03 月 01 日
52.  * 注    意:
53.  *********************************************************************************/
54.  void  InitCalcHour(void)
55.  {
56.  }
57.
58.  /*********************************************************************************
59.  * 函数名称: CalcHour
60.  * 函数功能: 计算小时
61.  * 输入参数: tick，即以秒为单位的计数值
62.  * 输出参数: void
63.  * 返 回 值: 小时值
64.  * 创建日期: 2021 年 03 月 01 日
65.  * 注    意:
66.  *********************************************************************************/
67.  short CalcHour(int tick)
68.  {
69.    short hour;
70.
71.    hour = tick / 3600;             //tick 对 3600 取模赋值给 hour
72.
73.    return(hour);
74.  }
```

步骤 4：完善 CalcMin.h 文件

将程序清单 9-3 中的代码输入 CalcMin.h 文件中。该文件声明了 2 个 API 函数，分别为 InitCalcMin 函数和 CalcMin 函数。

程序清单 9-3

```
1.   /*********************************************************************************
2.   * 模块名称: CalcMin.h
3.   * 摘    要: 计算分钟
4.   * 当前版本: 1.0.0
5.   * 作    者: Leyutek(COPYRIGHT 2018 - 2021 Leyutek. All rights reserved.)
6.   * 完成日期: 2021 年 03 月 01 日
7.   * 内    容:
8.   * 注    意:
9.   *********************************************************************************
10.  * 取代版本:
11.  * 作    者:
12.  * 完成日期:
13.  * 修改内容:
14.  * 修改文件:
15.  *********************************************************************************/
16.  #ifndef _CALC_MIN_H_
17.  #define _CALC_MIN_H_
```

```
18.
19. /*********************************************************************************************************
20. *                                              包含头文件
21. *********************************************************************************************************/
22.
23. /*********************************************************************************************************
24. *                                              宏定义
25. *********************************************************************************************************/
26.
27. /*********************************************************************************************************
28. *                                              枚举结构体定义
29. *********************************************************************************************************/
30.
31. /*********************************************************************************************************
32. *                                              API 函数声明
33. *********************************************************************************************************/
34. void  InitCalcMin(void);     //初始化 CalcMin 模块
35. short CalcMin(int tick);     //计算分钟值
36.
37. #endif
```

步骤 5：完善 CalcMin.c 文件

将程序清单 9-4 中的代码输入 CalcMin.c 文件中。该文件定义了 2 个 API 函数，分别为 InitCalcMin 函数和 CalcMin 函数。

程序清单 9-4

```
1.  /*********************************************************************************************************
2.  * 模块名称: CalcMin.c
3.  * 摘    要: 计算分钟
4.  * 当前版本: 1.0.0
5.  * 作    者: Leyutek(COPYRIGHT 2018 - 2021 Leyutek. All rights reserved.)
6.  * 完成日期: 2021 年 03 月 01 日
7.  * 内    容:
8.  * 注    意:
9.  **********************************************************************************************************
10. * 取代版本:
11. * 作    者:
12. * 完成日期:
13. * 修改内容:
14. * 修改文件:
15. *********************************************************************************************************/
16.
17. /*********************************************************************************************************
18. *                                              包含头文件
19. *********************************************************************************************************/
20. #include "CalcMin.h"
21.
22. /*********************************************************************************************************
23. *                                              宏定义
24. *********************************************************************************************************/
25.
26. /*********************************************************************************************************
```

```
27. *                                            枚举结构体定义
28. *********************************************************************************************/
29.
30. /*********************************************************************************************
31. *                                             内部变量
32. *********************************************************************************************/
33.
34. /*********************************************************************************************
35. *                                            内部函数声明
36. *********************************************************************************************/
37.
38. /*********************************************************************************************
39. *                                            内部函数实现
40. *********************************************************************************************/
41.
42. /*********************************************************************************************
43. *                                            API 函数实现
44. *********************************************************************************************/
45. /*********************************************************************************************
46. * 函数名称: InitCalcMin
47. * 函数功能: 初始化分钟计算模块
48. * 输入参数: void
49. * 输出参数: void
50. * 返 回 值: void
51. * 创建日期: 2021 年 03 月 01 日
52. * 注    意:
53. *********************************************************************************************/
54. void  InitCalcMin(void)
55. {
56. }
57.
58. /*********************************************************************************************
59. * 函数名称: CalcMin
60. * 函数功能: 计算分钟
61. * 输入参数: tick，即以秒为单位的计数值
62. * 输出参数: void
63. * 返 回 值: 分钟值
64. * 创建日期: 2021 年 03 月 01 日
65. * 注    意:
66. *********************************************************************************************/
67. short CalcMin(int tick)
68. {
69.   short min;
70.
71.   min = (tick % 3600) / 60;     //tick 对 3600 取余后再对 60 取模，赋值给 min
72.
73.   return(min);
74. }
```

步骤 6：完善 CalcSec.h 文件

将程序清单 9-5 中的代码输入 CalcSec.h 文件中。该文件声明了 2 个 API 函数，分别为 InitCalcSec 函数和 CalcSec 函数。

程序清单 9-5

```
1.  /***************************************************************************************
2.  * 模块名称: CalcSec.h
3.  * 摘    要: 计算秒
4.  * 当前版本: 1.0.0
5.  * 作    者: Leyutek(COPYRIGHT 2018 - 2021 Leyutek. All rights reserved.)
6.  * 完成日期: 2021 年 03 月 01 日
7.  * 内    容:
8.  * 注    意:
9.  ****************************************************************************************
10. * 取代版本:
11. * 作    者:
12. * 完成日期:
13. * 修改内容:
14. * 修改文件:
15. ***************************************************************************************/
16. #ifndef _CALC_SEC_H_
17. #define _CALC_SEC_H_
18.
19. /***************************************************************************************
20. *                                   包含头文件
21. ***************************************************************************************/
22.
23. /***************************************************************************************
24. *                                     宏定义
25. ***************************************************************************************/
26.
27. /***************************************************************************************
28. *                                  枚举结构体定义
29. ***************************************************************************************/
30.
31. /***************************************************************************************
32. *                                   API 函数声明
33. ***************************************************************************************/
34. void  InitCalcSec(void);    //初始化 CalcSec 模块
35. short CalcSec(int tick);    //计算秒值
36.
37. #endif
```

步骤 7：完善 CalcSec.c 文件

将程序清单 9-6 中的代码输入 CalcSec.c 文件中。该文件定义了 2 个 API 函数，分别为 InitCalcSec 函数和 CalcSec 函数。

程序清单 9-6

```
1.  /***************************************************************************************
2.  * 模块名称: CalcSec.c
3.  * 摘    要: 计算秒
4.  * 当前版本: 1.0.0
5.  * 作    者: Leyutek(COPYRIGHT 2018 - 2021 Leyutek. All rights reserved.)
6.  * 完成日期: 2021 年 03 月 01 日
7.  * 内    容:
8.  * 注    意:
```

```
9.  ****************************************************************************************
10. * 取代版本:
11. * 作    者:
12. * 完成日期:
13. * 修改内容:
14. * 修改文件:
15. ***************************************************************************************/
16.
17. /***************************************************************************************
18. *                                      包含头文件
19. ***************************************************************************************/
20. #include "CalcSec.h"
21.
22. /***************************************************************************************
23. *                                        宏定义
24. ***************************************************************************************/
25.
26. /***************************************************************************************
27. *                                    枚举结构体定义
28. ***************************************************************************************/
29.
30. /***************************************************************************************
31. *                                       内部变量
32. ***************************************************************************************/
33.
34. /***************************************************************************************
35. *                                     内部函数声明
36. ***************************************************************************************/
37.
38. /***************************************************************************************
39. *                                     内部函数实现
40. ***************************************************************************************/
41.
42. /***************************************************************************************
43. *                                     API 函数实现
44. ***************************************************************************************/
45. /***************************************************************************************
46. * 函数名称: InitCalcSec
47. * 函数功能: 初始化秒计算模块
48. * 输入参数: void
49. * 输出参数: void
50. * 返 回 值: void
51. * 创建日期: 2021 年 03 月 01 日
52. * 注    意:
53. ***************************************************************************************/
54. void  InitCalcSec(void)
55. {
56. }
57.
58. /***************************************************************************************
59. * 函数名称: CalcSec
60. * 函数功能: 计算秒
```

```
61. * 输入参数：tick，即以秒为单位的计数值
62. * 输出参数：void
63. * 返 回 值：秒值
64. * 创建日期：2021 年 03 月 01 日
65. * 注    意：
66. *********************************************************************************************/
67. short CalcSec(int tick)
68. {
69.   short sec;
70.
71.   sec = (tick % 3600) % 60;    //tick 对 3600 取余后再对 60 取余，赋值给 sec
72.
73.   return(sec);
74. }
```

步骤 8：完善 App.c 文件

将程序清单 9-7 中的代码输入 App.c 文件中。下面按照顺序对部分语句进行解释。

（1）第 23 至 25 行代码：在 App.c 文件的“包含头文件”区，包含 CalcHour、CalcMin 和 CalcSec 模块的头文件，这样就无须在 App.c 文件中声明这三个模块中的 API 函数，直接调用即可。

（2）第 68 至 70 行代码：在 main 函数中，通过调用 CalcHour、CalcMin 和 CalcSec 函数，计算小时、分钟和秒值。

程序清单 9-7

```
1.  /*********************************************************************************************
2.  * 模块名称：App.c
3.  * 摘    要：测试基于多文件的秒值-时间值转换实验
4.  * 当前版本：1.0.0
5.  * 作    者：Leyutek(COPYRIGHT 2018 - 2021 Leyutek. All rights reserved.)
6.  * 完成日期：2021 年 03 月 01 日
7.  * 内    容：
8.  * 注    意：
9.  *********************************************************************************************
10. * 取代版本：
11. * 作    者：
12. * 完成日期：
13. * 修改内容：
14. * 修改文件：
15. *********************************************************************************************/
16.
17. /*********************************************************************************************
18. *                                        包含头文件
19. *********************************************************************************************/
20. #include <stdio.h>
21. #include <stdlib.h>
22.
23. #include "CalcHour.h"
24. #include "CalcMin.h"
25. #include "CalcSec.h"
26.
27. /*********************************************************************************************
```

```
28. *                                          宏定义
29. *********************************************************************************/
30.
31. /*********************************************************************************
32. *                                       枚举结构体定义
33. *********************************************************************************/
34.
35.
/*********************************************************************************
**
36. *                                         内部变量
37. *********************************************************************************/
38.
39. /*********************************************************************************
40. *                                       内部函数声明
41. *********************************************************************************/
42.
43. /*********************************************************************************
44. *                                       内部函数实现
45. *********************************************************************************/
46.
47. /*********************************************************************************
48. *                                       API 函数实现
49. *********************************************************************************/
50. /*********************************************************************************
51. * 函数名称: main
52. * 函数功能: 主函数
53. * 输入参数: void
54. * 输出参数: void
55. * 返回值: void
56. * 创建日期: 2021 年 03 月 01 日
57. * 注    意:
58. *********************************************************************************/
59. void  main(void)
60. {
61.   int tick = 0;
62.
63.   short hour, min, sec;
64.
65.   printf("Please input a tick between 0-86399\n");
66.   scanf_s("%d", &tick);
67.
68.   hour = CalcHour(tick);
69.   min  = CalcMin(tick);
70.   sec  = CalcSec(tick);
71.
72.   printf("Current time: %02d-%02d-%02d\n", hour, min, sec);
73.
74.   system("pause");
75. }
```

步骤 9：项目编译和运行

最后，按 F5 键编译并运行程序，在弹出的控制台窗口中，输入 80000 后按回车键，可以看到运行结果，即输出“Current time: 22-13-20”，说明实验成功。

本 章 任 务

任务 1：2020 年有 366 天，将 2020 年 1 月 1 日作为计数起点，即计数 1，2020 年 12 月 31 日作为计数终点，即计数 366。计数 1 代表“2020 年 1 月 1 日-星期三”，计数 10 代表“2020 年 1 月 10 日-星期五”。参考本章实验，通过键盘输入一个 1～366 之间的值，包括 1 和 366，基于多文件，将其转换为年、月、日、星期，并输出到控制台窗口。另外，分别通过静态库和动态库的方式，实现 1～366 之间的值到年、月、日、星期的转换。

任务 2：参考本章实验，通过键盘输入 2 个数，判断大小，并输出结果到控制台窗口。

任务 3：参考本章实验，通过键盘输入 10 个数，判断大小，并输出结果到控制台窗口。

本 章 习 题

1．为什么要使用多文件？

2．什么是重编译？如何防止重编译？

3．简述内部函数和外部函数的区别。

第 10 章　基于多媒体定时器的电子钟设计

第 2～9 章实验通过不同的方法完成秒值-时间值转换，使读者对 C 语言有了初步的认识。本章和第 11 章将引入多媒体定时器的概念，在实现将秒值转换为时间值的基础上，进一步实现让秒值递增计数，每秒递增计数一次，从 0（对应的时间为 00:00:00）计数到 86399（对应的时间为 23:59:59），并通过 printf 函数每秒打印一次时间值。初始值可设置为 0～86399 之间的任意值。

10.1　实验内容

设计一个基于多媒体定时器的电子钟，在 App 模块中通过多媒体定时器实现 2ms 定时，并以此为秒值计数的时间基准，考虑到第 9 章已经实现的 CalcHour、CalcMin 和 CalcSec 模块复用，只需要新增两个模块，分别为 Tick 模块和 CalcTime 模块。其中，Tick 模块用于实现秒值的计数，CalcTime 模块分别调用 CalcHour、CalcMin 和 CalcSec 模块中的相关函数计算小时值、分钟值和秒值。最后，在 App 模块中，将动态的时间值通过控制台窗口输出，实现每秒打印一个完整的时间值。

10.2　实验原理

10.2.1　项目架构

图 10-1 为本章实验的项目架构，App 模块通过调用 Tick 模块实现秒值的计数、设定和获取，其中 InitTick 函数用于初始化 Tick 模块，TickPer2Ms 函数用于以秒为单位的时间值计数，SetTickVal 和 GetTickVal 函数分别用于设定和获取以秒为单位的时间值。App 模块通过调用 CalcTime 模块中的相关函数，将以秒值为单位的时间值转换为小时值、分钟值和秒值。此外，在 App 模块中实现 2ms 的多媒体定时器，以及打印时间值功能。

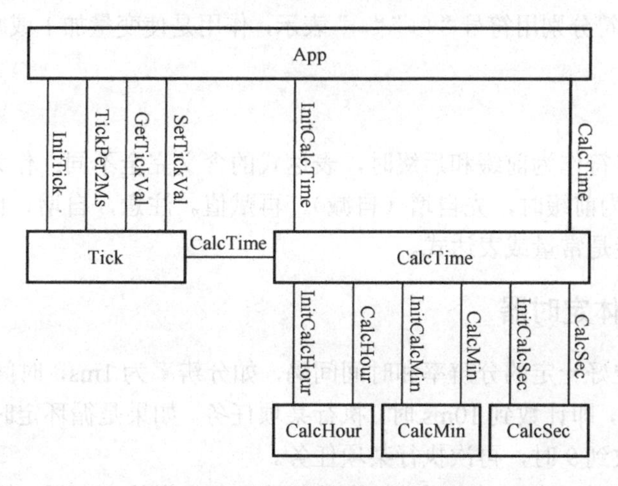

图 10-1　本章实验的项目架构

10.2.2　函数指针

指针除了可以指向变量地址，还可以指向函数，将指向函数的指针称为函数指针。

编译时，每个函数都有一个入口地址，该入口地址即为函数指针所指向的地址。如同利用指针变量可引用其他类型变量一样，利用函数指针也可以调用函数。函数指针有两个用途：调用函数、作为函数的参数。

10.2.3　回调函数

回调函数就是通过函数指针调用的函数。如果把函数的指针（地址）作为参数传递给另一个函数，当这个指针被用来调用其所指向的函数时，被调用的函数则称为回调函数。在特定的事件或条件发生时，回调函数由另一方调用，用于响应该事件或条件。

10.2.4　局部变量和全局变量

局部变量是在函数内部定义的变量，其作用域仅限于函数内部，在函数外部无法调用。全局变量则是定义于函数之外的变量，其作用域为整个源程序。

10.2.5　静态变量

静态变量的生命周期与源程序相同，定义方法是在类型名前加 static 关键字。

静态局部变量在函数内部定义，若在定义时未赋初值，则系统一般会自动为其赋值。非静态的局部变量在函数结束后就会消失，但静态局部变量在函数结束后仍然存在，只是在函数外部无法调用，作用域仅限于函数内部。

静态全局变量在函数外部定义，与非静态全局变量的区别是：当一个源程序由多个文件组成时，非静态全局变量的作用域为整个源程序，而静态全局变量的作用域仅限于定义该变量的源文件。

10.2.6　自增、自减运算符

自增、自减运算符分别用符号“++”“--”表示，作用是使变量加 1 或减 1。用法示例如下：

```
b = a++;
b = ++a;
```

自增、自减运算符作为前缀和后缀时，表达式的含义完全不同：作为后缀时，先赋值、再自增（自减）；作为前缀时，先自增（自减）、再赋值。注意，自增、自减运算符的作用对象只能是变量，不能是常量或表达式。

10.2.7　多媒体定时器

定时器需要设定好一定的分辨率和时间间隔，如分辨率为 1ms，时间间隔为 10ms，当系统从 0 开始计数到 9，即计数到 10ms 时，执行某项任务。如果是循环定时器，则定时器清零，继续开始计数，计数到 9 时，再次执行某项任务。

在 Windows 编程中，有两种定时器：普通定时器和多媒体定时器。普通定时器的精度低，而多媒体定时器的精度高，这里只介绍多媒体定时器。

多媒体定时器在使用时，先设定好延时周期、精度、回调函数、回调数据和定时器事件类型，定时器事件类型一般配置为周期性触发（TIME_PERIODIC）。下面用一个示例介绍如何使用多媒体定时器。

（1）在需要使用多媒体定时器模块的“头文件包含”区，包含"windows.h"和"winmm.lib"，具体代码如下：

```
#include <windows.h>
#pragma comment(lib, "winmm.lib")
```

（2）在“内部函数声明”区声明回调函数，如声明一个 TimeProc 的回调函数：

```
static void __stdcall TimeProc(unsigned int uTimerID, unsigned int uMsg, unsigned long dwUser,
unsigned long dw1, unsigned long dw2);
```

（3）在“内部函数实现”区，定义回调函数 TimeProc：

```
static void __stdcall TimeProc(unsigned int uTimerID, unsigned int uMsg, unsigned long dwUser,
unsigned long dw1, unsigned long dw2)
{
    //用户代码
}
```

（4）设置一个多媒体定时器事件，例如，设定一个延时周期为 2，精度为 1ms，回调函数为 TimeProc，回调数据为 0，定时器事件类型为周期性触发的多媒体定时器事件，代码如下：

```
timeSetEvent(2, 1, TimeProc, 0, TIME_PERIODIC);
```

10.3　实验步骤

步骤 1：复制原始项目

首先，将本书配套资料包的“03.例程资料\Material\09.基于多媒体定时器的电子钟设计实验”文件夹复制到 CProgramTest 文件夹中，然后，双击运行“D:\CProgramTest\09.基于多媒体定时器的电子钟设计实验\Project”文件夹中的 Project.sln 文件。

步骤 2：完善 CalcTime.h 文件

将程序清单 10-1 中的代码输入 CalcTime.h 文件中。该文件主要声明 CalcTime 模块的 2 个 API 函数，分别是初始化 CalcTime 模块的函数 InitCalcTime 和计算时间函数 CalcTime，其中，计算什么值通过参数传递不同的枚举元素来判断。

程序清单 10-1

```
1.   /*********************************************************************************
2.   * 模块名称: CalcTime.h
3.   * 摘    要: 秒值-时间值换算
4.   * 当前版本: 1.0.0
5.   * 作    者: Leyutek(COPYRIGHT 2018 - 2021 Leyutek. All rights reserved.)
6.   * 完成日期: 2021 年 03 月 01 日
7.   * 内    容:
8.   * 注    意:
9.   **********************************************************************************
10.  * 取代版本:
```

```
11.  * 作    者:
12.  * 完成日期:
13.  * 修改内容:
14.  * 修改文件:
15.  *********************************************************************************/
16.  #ifndef _CALC_TIME_H_
17.  #define _CALC_TIME_H_
18.
19.  /*********************************************************************************
20.  *                                   包含头文件
21.  *********************************************************************************/
22.
23.  /*********************************************************************************
24.  *                                     宏定义
25.  *********************************************************************************/
26.
27.  /*********************************************************************************
28.  *                                  枚举结构体定义
29.  *********************************************************************************/
30.  //定义枚举
31.  typedef enum
32.  {
33.    TIME_VAL_HOUR = 0,
34.    TIME_VAL_MIN,
35.    TIME_VAL_SEC,
36.    TIME_VAL_MAX
37.  }EnumTimeVal;
38.
39.  /*********************************************************************************
40.  *                                  API 函数声明
41.  *********************************************************************************/
42.  void  InitCalcTime(void);                             //初始化 CalcTime 模块
43.  short CalcTime(int tick, unsigned char type);         //计算时间
44.
45.  #endif
```

步骤 3：完善 CalcTime.c 文件

将程序清单 10-2 中的代码输入 CalcTime.c 文件中。该文件主要是定义 CalcTime 模块的 2 个 API 函数。

程序清单 10-2

```
1.   /*********************************************************************************
2.   * 模块名称: CalcTime.c
3.   * 摘    要: 秒值-时间值换算
4.   * 当前版本: 1.0.0
5.   * 作    者: Leyutek(COPYRIGHT 2018 - 2021 Leyutek. All rights reserved.)
6.   * 完成日期: 2021 年 03 月 01 日
7.   * 内    容:
8.   * 注    意:
9.   *********************************************************************************
10.  * 取代版本:
11.  * 作    者:
```

```
12. * 完成日期:
13. * 修改内容:
14. * 修改文件:
15. *********************************************************************************/
16.
17. /*********************************************************************************
18. *                                   包含头文件
19. *********************************************************************************/
20. #include "CalcHour.h"
21. #include "CalcMin.h"
22. #include "CalcSec.h"
23. #include "CalcTime.h"
24.
25. /*********************************************************************************
26. *                                     宏定义
27. *********************************************************************************/
28.
29. /*********************************************************************************
30. *                                  枚举结构体定义
31. *********************************************************************************/
32.
33. /*********************************************************************************
34. *                                    内部变量
35. *********************************************************************************/
36.
37. /*********************************************************************************
38. *                                  内部函数声明
39. *********************************************************************************/
40.
41. /*********************************************************************************
42. *                                  内部函数实现
43. *********************************************************************************/
44.
45. /*********************************************************************************
46. *                                  API 函数实现
47. *********************************************************************************/
48. /*********************************************************************************
49. * 函数名称: InitCalcTime
50. * 函数功能: 初始化 CalcTime 模块
51. * 输入参数: void
52. * 输出参数: void
53. * 返 回 值: void
54. * 创建日期: 2021 年 03 月 01 日
55. * 注    意:
56. *********************************************************************************/
57. void  InitCalcTime(void)
58. {
59.   InitCalcHour();
60.   InitCalcMin();
61.   InitCalcSec();
62. }
63.
```

```
64.  /*****************************************************************************************
65.  * 函数名称: CalcTime
66.  * 函数功能: 计算时间
67.  * 输入参数: tick，即以秒为单位的计数值
68.  * 输出参数: void
69.  * 返 回 值: hour、min 或 sec，具体是哪一个根据 type 来确定
70.  * 创建日期: 2021 年 03 月 01 日
71.  * 注    意:
72.  *****************************************************************************************/
73.  short CalcTime(int tick, unsigned char type)
74.  {
75.    short timeVal;
76.
77.    switch(type)
78.    {
79.      case TIME_VAL_HOUR:
80.        timeVal = CalcHour(tick);    //通过调用 CalcHour 计算小时值
81.        break;
82.      case TIME_VAL_MIN:
83.        timeVal = CalcMin(tick);     //通过调用 CalcMin 计算分钟值
84.        break;
85.      case TIME_VAL_SEC:
86.        timeVal = CalcSec(tick);     //通过调用 CalcSec 计算秒值
87.        break;
88.      default:
89.        break;
90.    }
91.
92.    return(timeVal);
93.  }
```

步骤 4：完善 Tick.h 文件

将程序清单 10-3 中的代码输入 Tick.h 文件中。该文件主要用于声明 Tick 模块的 4 个 API 函数。

程序清单 10-3

```
1.   /*****************************************************************************************
2.   * 模块名称: Tick.h
3.   * 摘    要: 时间滴答计数，以及时间值保存和获取
4.   * 当前版本: 1.0.0
5.   * 作    者: Leyutek(COPYRIGHT 2018 - 2021 Leyutek. All rights reserved.)
6.   * 完成日期: 2021 年 03 月 01 日
7.   * 内    容:
8.   * 注    意:
9.   *****************************************************************************************
10.  * 取代版本:
11.  * 作    者:
12.  * 完成日期:
13.  * 修改内容:
14.  * 修改文件:
15.  *****************************************************************************************/
16.  #ifndef _TICK_H_
```

```
17.  #define _TICK_H_
18.
19.  /*********************************************************************************
20.  *                                   包含头文件
21.  *********************************************************************************/
22.
23.  /*********************************************************************************
24.  *                                   宏定义
25.  *********************************************************************************/
26.
27.  /*********************************************************************************
28.  *                                   枚举结构体定义
29.  *********************************************************************************/
30.
31.  /*********************************************************************************
32.  *                                   API 函数声明
33.  *********************************************************************************/
34.  void  InitTick(void);                    //初始化 Tick 模块
35.  void  TickPer2Ms(void);                  //每 2ms 执行一次
36.
37.  void  SetTickVal(int tick);              //设置 tick 值
38.  int   GetTickVal(void);                  //获取 tick 值
39.
40.  #endif
```

步骤 5：完善 Tick.c 文件

将程序清单 10-4 中的代码输入 Tick.c 文件中。该文件主要用于定义 Tick 模块的 4 个 API 函数。

（1）第 60 至 89 行代码：TickPer2Ms 函数每 2ms 执行一次，变量 s_iCnt 循环递增计数，计数范围为 0～499，当计数到 499 时，执行第 75 至 83 行代码，即每秒变量 s_iTickVal 递增计数，计数范围为 0～86399，并将 s_iCnt 清零。

（2）第 91 至 117 行代码：SetTickVal 函数用于设置变量 s_iTickVal，GetTickVal 函数用于获取变量 s_iTickVal，这两个函数通常由其他文件中的函数调用，这样就可以实现变量的文件间读写操作，而无须使用全局变量的方式，当然，也不建议使用全局变量。注意，如果是在定义变量 s_iTickVal 的文件中对该变量进行读写操作，则无须使用这两个函数。

程序清单 10-4

```
1.   /*********************************************************************************
2.   * 模块名称：Tick.c
3.   * 摘    要：时间滴答计数，以及时间值保存和获取
4.   * 当前版本：1.0.0
5.   * 作    者：Leyutek(COPYRIGHT 2018 - 2021 Leyutek. All rights reserved.)
6.   * 完成日期：2021 年 03 月 01 日
7.   * 内    容：
8.   * 注    意：
9.   *********************************************************************************
10.  * 取代版本：
11.  * 作    者：
12.  * 完成日期：
13.  * 修改内容：
```

```
14. * 修改文件:
15. ********************************************************************************/
16.
17. /*********************************************************************************
18. *                                  包含头文件
19. ********************************************************************************/
20. #include "Tick.h"
21.
22. /*********************************************************************************
23. *                                   宏定义
24. ********************************************************************************/
25.
26. /*********************************************************************************
27. *                                枚举结构体定义
28. ********************************************************************************/
29.
30. /*********************************************************************************
31. *                                  内部变量
32. ********************************************************************************/
33. static  int s_iTickVal;
34.
35. /*********************************************************************************
36. *                                内部函数声明
37. ********************************************************************************/
38.
39. /*********************************************************************************
40. *                                内部函数实现
41. ********************************************************************************/
42.
43. /*********************************************************************************
44. *                                API 函数实现
45. ********************************************************************************/
46. /*********************************************************************************
47. * 函数名称: InitTick
48. * 函数功能: 初始化 Tick 模块
49. * 输入参数: void
50. * 输出参数: void
51. * 返 回 值: void
52. * 创建日期: 2021 年 03 月 01 日
53. * 注    意:
54. ********************************************************************************/
55. void  InitTick(void)
56. {
57.   s_iTickVal = 0;
58. }
59.
60. /*********************************************************************************
61. * 函数名称: TickPer2Ms
62. * 函数功能: 每 2ms 执行一次
63. * 输入参数: void
64. * 输出参数: void
65. * 返 回 值: void
```

```
66. * 创建日期: 2021 年 03 月 01 日
67. * 注    意:
68. ********************************************************************************/
69. void  TickPer2Ms(void)
70. {
71.   static  short s_iCnt = 0;
72.
73.   if(s_iCnt >= 499)
74.   {
75.     if(s_iTickVal >= 86399)
76.     {
77.       s_iTickVal = 0;
78.     }
79.     else
80.     {
81.       s_iTickVal++;
82.     }
83.     s_iCnt = 0;
84.   }
85.   else
86.   {
87.     s_iCnt++;
88.   }
89. }
90.
91. /*********************************************************************************
92. * 函数名称: SetTickVal
93. * 函数功能: 设置 tick 值
94. * 输入参数: tick
95. * 输出参数: void
96. * 返 回 值: void
97. * 创建日期: 2021 年 03 月 01 日
98. * 注    意:
99. ********************************************************************************/
100. void  SetTickVal(int tick)
101. {
102.   s_iTickVal = tick;
103. }
104.
105. /*********************************************************************************
106. * 函数名称: GetTickVal
107. * 函数功能: 获取 tick 值
108. * 输入参数: void
109. * 输出参数: void
110. * 返 回 值: tick 值
111. * 创建日期: 2021 年 03 月 01 日
112. * 注    意:
113. ********************************************************************************/
114. int   GetTickVal(void)
115. {
116.   return(s_iTickVal);
117. }
```

步骤 6：完善 App.c 文件

将程序清单 10-5 中的代码输入 App.c 文件中。下面按照顺序对部分语句进行解释。

（1）第 53 至 68 行代码：InitSoftware 函数主要用于初始化软件模块，即执行各个软件模块的初始化函数。通过 SetTickVal 函数设置初始化时间。注意，在嵌入式系统程序设计中，与 InitSoftware 函数对应的 InitHardware 函数用于初始化硬件相关的模块，如 InitUART、InitTimer 等。

（2）第 70 至 83 行代码：TimeProc 是定时器的回调函数，该函数的函数体只有一行代码，因此，每次触发定时器时，实际上执行的是 Proc2msTask 函数。

（3）第 85 至 109 行代码：每执行一次 Proc2msTask 函数，变量 s_iCnt500 循环递增计数，计数范围为 0～499，当计数到 499 时，执行 Proc1SecTask 函数。Proc2msTask 函数每 2ms 执行一次，因此，TickPer2Ms 函数同样每 2ms 执行一次，Proc1SecTask 函数每 1s 执行一次。

（4）第 111 至 133 行代码：Proc1SecTask 函数的功能主要是获取以秒为单位的时间值，然后将其转换为小时值、分钟值和秒值，并通过 printf 函数打印时间值。

（5）第 161 行代码：timeSetEvent 函数用于启动一个定时器，该定时器的精度为 1ms，每 2ms 触发一次定时器，并执行回调函数 TimeProc。

程序清单 10-5

```
1.  /*********************************************************************************************
2.  * 模块名称: App.c
3.  * 摘    要: 测试基于多媒体定时器的电子钟设计实验
4.  * 当前版本: 1.0.0
5.  * 作    者: Leyutek(COPYRIGHT 2018 - 2021 Leyutek. All rights reserved.)
6.  * 完成日期: 2021 年 03 月 01 日
7.  * 内    容:
8.  * 注    意:
9.  *********************************************************************************************
10. * 取代版本:
11. * 作    者:
12. * 完成日期:
13. * 修改内容:
14. * 修改文件:
15. *********************************************************************************************/
16.
17. /*********************************************************************************************
18. *                                          包含头文件
19. *********************************************************************************************/
20. #include <stdio.h>
21. #include <windows.h>
22. #pragma comment(lib, "winmm.lib")    //导入 winmm.lib 多媒体库
23.
24. #include "CalcTime.h"
25. #include "Tick.h"
26.
27. /*********************************************************************************************
28. *                                           宏定义
29. *********************************************************************************************/
30.
31. /*********************************************************************************************
```

```
32. *                                        枚举结构体定义
33. *********************************************************************************/
34.
35. /*********************************************************************************
36. *                                        内部变量
37. *********************************************************************************/
38.
39. /*********************************************************************************
40. *                                        内部函数声明
41. *********************************************************************************/
42. static  void  InitSoftware(void);      //初始化 Software
43.
44. static void __stdcall TimeProc(unsigned int uTimerID, unsigned int uMsg, unsigned long dwUser,
45.   unsigned long dw1, unsigned long dw2);   //定时器回调函数
46.
47. static  void  Proc2msTask(void);        //声明一个 2ms 执行一次的函数
48. static  void  Proc1SecTask(void);       //声明一个 1s 执行一次的函数
49.
50. /*********************************************************************************
51. *                                        内部函数实现
52. *********************************************************************************/
53. /*********************************************************************************
54. * 函数名称: InitSoftware
55. * 函数功能: 所有的软件初始化函数都放在此函数中
56. * 输入参数: void
57. * 输出参数: void
58. * 返 回 值: void
59. * 创建日期: 2021 年 03 月 01 日
60. * 注    意:
61. *********************************************************************************/
62. static  void  InitSoftware(void)
63. {
64.   InitCalcTime();
65.   InitTick();
66.
67.   SetTickVal(80000);                  //设定初始化时间
68. }
69.
70. /*********************************************************************************
71. * 函数名称: TimeProc
72. * 函数功能: 定时器回调函数
73. * 输入参数: uTimerID, uMsg, dwUser, dw1, dw2, 详见 MSDN
74. * 输出参数: void
75. * 返 回 值: void
76. * 创建日期: 2021 年 03 月 01 日
77. * 注    意:
78. *********************************************************************************/
79. static void __stdcall TimeProc(unsigned int uTimerID, unsigned int uMsg, unsigned long dwUser,
80.   unsigned long dw1, unsigned long dw2)
81. {
82.   Proc2msTask();
83. }
```

```
84.
85. /******************************************************************************
86. * 函数名称: Proc2msTask
87. * 函数功能: 处理 2ms 任务
88. * 输入参数: void
89. * 输出参数: void
90. * 返 回 值: void
91. * 创建日期: 2021 年 03 月 01 日
92. * 注    意:
93. ******************************************************************************/
94. static  void  Proc2msTask(void)
95. {
96.   static  short s_iCnt500 = 0;
97.
98.   TickPer2Ms();
99.
100.  if(s_iCnt500 >=499)
101.  {
102.    Proc1SecTask();
103.    s_iCnt500 = 0;
104.  }
105.  else
106.  {
107.    s_iCnt500++;
108.  }
109. }
110.
111. /*****************************************************************************
112. * 函数名称: Proc1SecTask
113. * 函数功能: 处理 1s 任务
114. * 输入参数: void
115. * 输出参数: void
116. * 返 回 值: void
117. * 创建日期: 2021 年 03 月 01 日
118. * 注    意:
119. *****************************************************************************/
120. static  void  Proc1SecTask(void)
121. {
122.  int tickVal;
123.  short hour, min, sec;
124.
125.  tickVal = GetTickVal();
126.
127.  hour = CalcTime(tickVal, TIME_VAL_HOUR);
128.  min  = CalcTime(tickVal, TIME_VAL_MIN);
129.  sec  = CalcTime(tickVal, TIME_VAL_SEC);
130.
131.  printf("Current time: %02d-%02d-%02d\n", hour, min, sec);
132.
133. }
134.
135. /*****************************************************************************
```

```
136. *                                                   API 函数实现
137. *********************************************************************************************/
138. /*********************************************************************************************
139. * 函数名称: main
140. * 函数功能: 主函数
141. * 输入参数: void
142. * 输出参数: void
143. * 返 回 值: void
144. * 创建日期: 2021 年 03 月 01 日
145. * 注    意:
146. *********************************************************************************************/
147. void  main(void)
148. {
149.   InitSoftware();                              //初始化软件
150.
151.   //timeSetEvent 函数说明
152.   //MMRESULT timeSetEvent(
153.   //  UINT           uDelay,           //以 ms 指定事件的周期
154.   //  UINT           uResolution,      //以 ms 指定延时的精度，数值越小定时器事件分辨率越高，
                                                                                   默认值为 1ms
155.   //  LPTIMECALLBACK lpTimeProc,       //指向一个回调函数，即单次事件或周期性事件触发时调用
                                                                                   的函数
156.   //  DWORD_PTR      dwUser,           //存放用户提供的回调数据
157.   //  UINT           fuEvent           //指定定时器事件类型，      TIME_ONESHOT-单次触发，
                                                                   TIME_PERIODIC-周期性触发
158.   //);
159.
160.   //用户定时器设定，定时器精度为 1ms，每隔 2ms 触发一次定时器，并执行回调函数 TimeProc
161.   timeSetEvent(2, 1, TimeProc, 0, TIME_PERIODIC);
162.
163.   while(1)
164.   {
165.   }
166. }
```

步骤 7：项目编译和运行

最后，按 F5 键编译并运行程序，在弹出的控制台窗口中，可以看到如图 10-2 所示的运行结果，即每秒打印一个完整的时间值，且时间值每打印一次递增 1s，说明实验成功。

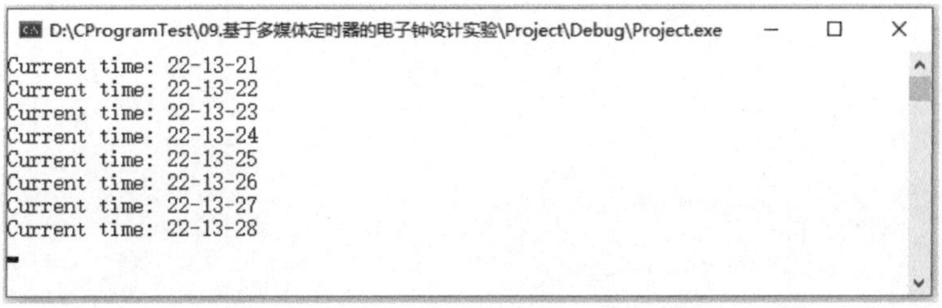

```
D:\CProgramTest\09.基于多媒体定时器的电子钟设计实验\Project\Debug\Project.exe
Current time: 22-13-21
Current time: 22-13-22
Current time: 22-13-23
Current time: 22-13-24
Current time: 22-13-25
Current time: 22-13-26
Current time: 22-13-27
Current time: 22-13-28
```

图 10-2　本章项目运行结果

本 章 任 务

2020 年有 366 天，将 2020 年 1 月 1 日作为计数起点，即计数 1，2020 年 12 月 31 日作为计数终点，即计数 366。计数 1 代表“2020 年 1 月 1 日-星期三”，计数 10 代表“2020 年 1 月 10 日-星期五”。参考本章实验，设计一个实验，实现每秒计数递增一次，计数范围为 1～366，并通过 printf 函数每秒打印一次计数对应的年、月、日、星期。例如，初始日期设置为 10，即“2020 年 1 月 10 日-星期五”，则第 1 秒打印“2020 年 1 月 10 日-星期五”，第 2 秒打印“2020 年 1 月 11 日-星期六”，以此类推。

本 章 习 题

1．本实验为什么要导入 winmm.lib 库？

2．什么是回调函数？什么是函数指针？

3．多媒体定时器的 timeSetEvent 函数都有哪些参数？每个参数代表什么意义？

第11章 电子钟的API设计与应用

本章实验实现的功能是每秒计数一次，从0（对应的时间为00:00:00）计数到86399（对应的时间为23:59:59），并通过 printf 函数每秒打印一次时间，初始值可以设置为0～86399中的任意值。

11.1 实验内容

本章实验在第10章实验的基础上，将 Tick、CalcTime、CalcHour、CalcMin 和 CalcSec 模块整合为 RunClock 模块。在 App 模块中通过多媒体定时器实现 2ms 定时功能，以此为秒值计数的时间基准。RunClock 模块中的 InitRunClock 函数用于实现模块的初始化，RunClockPer2Ms 函数用于计数，SetTimeVal 函数用于设置时间值，GetTimeVal 函数用于获取时间值，PauseClock 函数用于实现电子钟的启动和暂停，DispTime 函数用于显示时间，App 模块通过调用 RunClock 模块中的函数来实现时钟的运行，以及控制台窗口每秒输出一个完整的时间值。

11.2 实验原理

11.2.1 项目架构

图11-1为本章实验的项目架构，各函数的功能如“实验内容”部分所述。

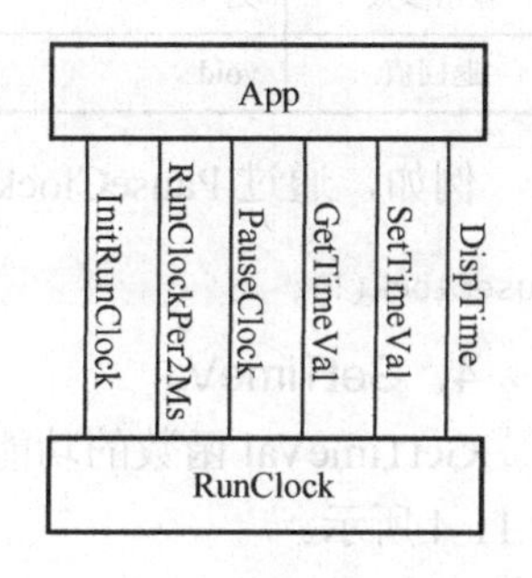

图11-1 本章实验项目架构

11.2.2 RunClock 模块函数

RunClock 模块由 RunClock.h 和 RunClock.c 文件实现。RunClock 模块中有 6 个 API 函数，分别为 InitRunClock、RunClockPer2Ms、PauseClock、GetTimeVal、SetTimeVal 和 DispTime，下面对这 6 个 API 函数进行介绍。

1. InitRunClock

InitRunClock 函数的功能是初始化 RunClock 模块，通过对 s_iHour、s_iMin 和 s_iSec 这 3 个内部变量赋值 0 来实现。该函数的描述如表11-1所示。

表11-1 InitRunClock 函数的描述

函数名	InitRunClock
函数原型	void InitRunClock(void)
功能描述	初始化 RunClock 模块
输入参数	void
输出参数	无
返回值	void

2. RunClockPer2Ms

RunClockPer2Ms 函数的功能是以 2ms 为最小单位运行时钟系统，该函数每执行 500 次，变量 s_iSec 递增一次。该函数的描述如表 11-2 所示。

表 11-2　RunClockPer2Ms 函数的描述

函数名	RunClockPer2Ms
函数原型	void RunClockPer2Ms(void)
功能描述	时钟计数，每 2ms 调用一次
输入参数	void
输出参数	无
返回值	void

3. PauseClock

PauseClock 函数的功能是启动和暂停时钟。该函数的描述如表 11-3 所示。

表 11-3　PauseClock 函数的描述

函数名	PauseClock
函数原型	void PauseClock(u8 flag)
功能描述	实现时钟的启动和暂停
输入参数	flag：时钟启动或暂停标志位。1-暂停时钟；0-启动时钟
输出参数	无
返回值	void

例如，通过 PauseClock 函数暂停时钟运行的代码如下：

```
PauseClock(1);
```

4. GetTimeVal

GetTimeVal 函数的功能是获取当前时间值，时间值的类型由 type 决定。该函数的描述如表 11-4 所示。

表 11-4　GetTimeVal 函数的描述

函数名	GetTimeVal
函数原型	i16 GetTimeVal(u8 type)
功能描述	获取当前的时间值
输入参数	type：时间值的类型
输出参数	无
返回值	获取到的当前时间值，如小时、分钟或秒值，类型由参数 type 决定

例如，通过 GetTimeVal 函数获取当前时间值的代码如下：

```
u8 hour;
u8 min;
u8 sec;
hour = GetTimeVal(TIME_VAL_HOUR);
```

```
min = GetTimeVal(TIME_VAL_MIN);
sec = GetTimeVal(TIME_VAL_SEC);
```

5. SetTimeVal

SetTimeVal 函数的功能是根据参数 timeVal 设置当前的时间值，时间值的类型由 type 决定。该函数的描述如表 11-5 所示。

表 11-5　SetTimeVal 函数的描述

函数名	SetTimeVal
函数原型	void SetTimeVal(u8 type, i16 timeVal)
功能描述	设置当前的时间值
输入参数	type：时间值的类型；timeVal：要设置的时间值类型
输出参数	无
返回值	void

例如，通过 SetTimeVal 函数将当前时间设置为 23:59:50，代码如下：

```
SetTimeVal(TIME_VAL_HOUR, 23);
SetTimeVal(TIME_VAL_MIN, 59);
SetTimeVal(TIME_VAL_SEC, 50);
```

6. DispTime

DispTime 函数的功能是根据参数 hour、min 和 sec 显示当前的时间，通过 printf 函数来实现。该函数的描述如表 11-6 所示。

表 11-6　DispTime 函数的描述

函数名	DispTime
函数原型	void DispTime(i16 hour, i16 min, i16 sec)
功能描述	显示当前的时间
输入参数	hour：当前的小时值；min：当前的分钟值；sec：当前的秒值
输出参数	无
返回值	void

例如，当前时间是 23:59:50，通过 DispTime 函数显示当前时间，代码如下：

```
DispTime(23, 59, 50);
```

11.2.3　DataType.h

在编写代码时，为了提高代码的输入效率，可将一些较长的关键字用缩写代替，例如，unsigned char 用 u8 代替。在 DataType.h 中，就使用新的数据类型名来代替已有的数据类型名，如下所示：

```
typedef signed char         i8;
typedef signed short        i16;
typedef signed int          i32;
typedef unsigned char       u8;
typedef unsigned short      u16;
typedef unsigned int        u32;
```

另外，由于布尔型常量也常在程序中用到，这里在 DataType.h 中增加了 TRUE 和 FALSE 的定义，如下所示：

```
#define TRUE        1
#define FALSE       0
```

11.2.4　while 循环语句

在 C 语言程序中，while 语句是常用的实现循环结构的语句，其一般形式如下：

```
while(表达式)
{
    循环体语句;
}
```

表达式为循环的判定条件，若表达式的值为真，则执行循环体语句；若为假，则跳出 while 循环。

while 循环语句的流程图如图 11-2 所示。

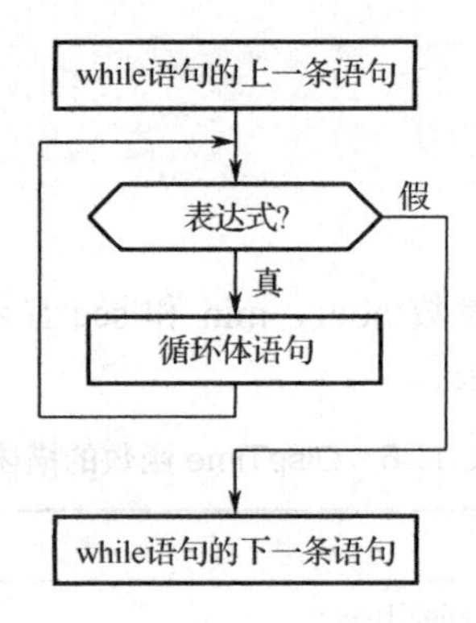

图 11-2　while 循环语句的流程图

11.3　实验步骤

步骤 1：复制原始项目

首先，将本书配套资料包的“03.例程资料\Material\10.电子钟的 API 设计与应用实验”文件夹复制到 CProgramTest 文件夹中，然后，双击运行“D:\CProgramTest\10.电子钟的 API 设计与应用实验\Project”文件夹中的 Project.sln 文件。

步骤 2：完善 RunClock.h 文件

将程序清单 11-1 中的代码输入 RunClock.h 文件中。该文件主要声明 RunClock 模块中的 6 个 API 函数。

程序清单 11-1

```
1.  /*********************************************************************************************
2.  * 模块名称: RunClock.h
3.  * 摘    要: 模拟时钟运行
4.  * 当前版本: 1.0.0
5.  * 作    者: Leyutek(COPYRIGHT 2018 - 2021 Leyutek. All rights reserved.)
6.  * 完成日期: 2021 年 03 月 01 日
7.  * 内    容:
```

```
8.  * 注    意:
9.  ***********************************************************************************
10. * 取代版本:
11. * 作    者:
12. * 完成日期:
13. * 修改内容:
14. * 修改文件:
15. ***********************************************************************************/
16. #ifndef _RUN_CLOCK_H_
17. #define _RUN_CLOCK_H_
18.
19. /**********************************************************************************
20. *                                   包含头文件
21. ***********************************************************************************/
22. #include "DataType.h"
23.
24. /**********************************************************************************
25. *                                     宏定义
26. ***********************************************************************************/
27.
28. /**********************************************************************************
29. *                                  枚举结构体定义
30. ***********************************************************************************/
31. //定义枚举
32. typedef enum
33. {
34.   TIME_VAL_HOUR = 0,
35.   TIME_VAL_MIN,
36.   TIME_VAL_SEC,
37.   TIME_VAL_MAX
38. }EnumTimeVal;
39.
40. /**********************************************************************************
41. *                                  API 函数声明
42. ***********************************************************************************/
43. void  InitRunClock(void);                  //初始化 RunClock 模块
44.
45. void  RunClockPer2Ms(void);                //每 2ms 调用一次
46. void  PauseClock(u8 flag);                 //flag，TRUE-暂停，FALSE-正常运行
47.
48. i16   GetTimeVal(u8 type);                 //获取时间值
49. void  SetTimeVal(u8 type, i16 timeVal);    //设置时间值
50.
51. void  DispTime(i16 hour, i16 min, i16 sec); //显示当前的时间
52.
53. #endif
```

步骤 3：完善 RunClock.c 文件

将程序清单 11-2 中的代码输入 RunClock.c 文件中。该文件主要定义 RunClock 模块中的 6 个 API 函数，这些函数的描述可参见 11.2.2 节，这里不再解释。

程序清单 11-2

```
1.  /*********************************************************************************************
2.  * 模块名称: RunClock.c
3.  * 摘    要: 模拟时钟运行
4.  * 当前版本: 1.0.0
5.  * 作    者: Leyutek(COPYRIGHT 2018 - 2021 Leyutek. All rights reserved.)
6.  * 完成日期: 2021 年 03 月 01 日
7.  * 内    容:
8.  * 注    意:
9.  **********************************************************************************************
10. * 取代版本:
11. * 作    者:
12. * 完成日期:
13. * 修改内容:
14. * 修改文件:
15. *********************************************************************************************/
16.
17. /*********************************************************************************************
18. *                                          包含头文件
19. *********************************************************************************************/
20. #include "RunClock.h"
21. #include <stdio.h>
22.
23. /*********************************************************************************************
24. *                                          宏定义
25. *********************************************************************************************/
26.
27. /*********************************************************************************************
28. *                                          枚举结构体定义
29. *********************************************************************************************/
30.
31. /*********************************************************************************************
32. *                                          内部变量
33. *********************************************************************************************/
34. static  i16 s_iHour;
35. static  i16 s_iMin ;
36. static  i16 s_iSec ;
37.
38. static  u8  s_iClockPauseFlag = 0;  //TRUE-暂停，FALSE-正常运行
39.
40. /*********************************************************************************************
41. *                                          内部函数声明
42. *********************************************************************************************/
43.
44. /*********************************************************************************************
45. *                                          内部函数实现
46. *********************************************************************************************/
47. /*********************************************************************************************
48. *                                          API 函数实现
49. *********************************************************************************************/
50. /*********************************************************************************************
51. * 函数名称: InitRunClock
```

```
52.  * 函数功能: 初始化模块
53.  * 输入参数: void
54.  * 输出参数: void
55.  * 返 回 值: void
56.  * 创建日期: 2021 年 03 月 01 日
57.  * 注    意:
58.  *********************************************************************************/
59.  void  InitRunClock(void)
60.  {
61.    s_iHour = 0;
62.    s_iMin  = 0;
63.    s_iSec  = 0;
64.  }
65.
66.  /*********************************************************************************
67.  * 函数名称: RunClockPer2Ms
68.  * 函数功能: 计数模块，每 2ms 调用一次
69.  * 输入参数: void
70.  * 输出参数: void
71.  * 返 回 值: void
72.  * 创建日期: 2021 年 03 月 01 日
73.  * 注    意:
74.  *********************************************************************************/
75.  void  RunClockPer2Ms(void)
76.  {
77.    static i16 s_iCnt500 = 0;
78.
79.    if(499 <= s_iCnt500 && 0 == s_iClockPauseFlag)
80.    {
81.      if(59 <= s_iSec)
82.      {
83.        if(59 <= s_iMin)
84.        {
85.          if(23 <= s_iHour)
86.          {
87.            s_iHour = 0;
88.          }
89.          else
90.          {
91.            s_iHour++;
92.          }
93.          s_iMin = 0;
94.        }
95.        else
96.        {
97.          s_iMin++;
98.        }
99.        s_iSec = 0;
100.     }
101.     else
102.     {
103.       s_iSec++;
```

```
104.     }
105.     s_iCnt500 = 0;
106.   }
107.   else
108.   {
109.     s_iCnt500++;
110.   }
111. }
112.
113. /*********************************************************************************************
114. * 函数名称: PauseClock
115. * 函数功能: 实现时钟的启动和暂停
116. * 输入参数: flag-1-Pause, 0-Running
117. * 输出参数: void
118. * 返 回 值: void
119. * 创建日期: 2021 年 03 月 01 日
120. * 注    意:
121. *********************************************************************************************/
122. void  PauseClock(u8 flag)
123. {
124.   s_iClockPauseFlag = flag;
125. }
126.
127. /*********************************************************************************************
128. * 函数名称: GetTimeVal
129. * 函数功能: 获取当前的时间值
130. * 输入参数: 时间值的类型
131. * 输出参数: void
132. * 返 回 值: 获取的时间值，小时、分钟或秒
133. * 创建日期: 2021 年 03 月 01 日
134. * 注    意:
135. *********************************************************************************************/
136. i16   GetTimeVal(u8 type)
137. {
138.   i16 timeVal;
139.
140.   switch(type)
141.   {
142.   case TIME_VAL_HOUR:
143.     timeVal = s_iHour;
144.     break;
145.   case TIME_VAL_MIN:
146.     timeVal = s_iMin;
147.     break;
148.   case TIME_VAL_SEC:
149.     timeVal = s_iSec;
150.     break;
151.   default:
152.     break;
153.   }
154.
155.   return(timeVal);
```

```
156. }
157.
158. /*********************************************************************************
159. * 函数名称: SetTimeVal
160. * 函数功能: 设置当前的时间值
161. * 输入参数: type-时间值的类型, timeVal-要设置的时间值
162. * 输出参数: void
163. * 返 回 值: void
164. * 创建日期: 2021 年 03 月 01 日
165. * 注     意:
166. *********************************************************************************/
167. void  SetTimeVal(u8 type, i16 timeVal)
168. {
169.   switch(type)
170.   {
171.     case TIME_VAL_HOUR:
172.       s_iHour = timeVal;
173.       break;
174.     case TIME_VAL_MIN:
175.       s_iMin  = timeVal;
176.       break;
177.     case TIME_VAL_SEC:
178.       s_iSec  = timeVal;
179.       break;
180.     default:
181.       break;
182.   }
183. }
184.
185. /*********************************************************************************
186. * 函数名称: DispTime
187. * 函数功能: 显示当前的时间
188. * 输入参数: hour, min, sec
189. * 输出参数: void
190. * 返 回 值: void
191. * 创建日期: 2021 年 03 月 01 日
192. * 注     意:
193. *********************************************************************************/
194. void  DispTime(i16 hour, i16 min, i16 sec)    //显示当前的时间
195. {
196.   printf("Now is %02d:%02d:%02d\n", hour, min, sec);
197. }
```

步骤 4：完善 App.c 文件

将程序清单 11-3 中的代码输入 App.c 文件中。该文件与第 10 章的 App.c 文件类似，这里不再解释。

程序清单 11-3

```
1.    /*********************************************************************************
2.    * 模块名称: App.c
3.    * 摘     要: 测试时钟
4.    * 当前版本: 1.0.0
```

```
5.  * 作    者: Leyutek(COPYRIGHT 2018 - 2021 Leyutek. All rights reserved.)
6.  * 完成日期: 2021 年 03 月 01 日
7.  * 内    容:
8.  * 注    意:
9.  *********************************************************************************
10. * 取代版本:
11. * 作    者:
12. * 完成日期:
13. * 修改内容:
14. * 修改文件:
15. *********************************************************************************/
16.
17. /********************************************************************************
18. *                                   包含头文件
19. *********************************************************************************/
20. #include <stdio.h>
21. #include <windows.h>
22. #pragma comment(lib, "winmm.lib")    //导入 winmm.lib 多媒体库
23.
24. #include "RunClock.h"
25. #include "DataType.h"
26.
27. /********************************************************************************
28. *                                    宏定义
29. *********************************************************************************/
30.
31. /********************************************************************************
32. *                                 枚举结构体定义
33. *********************************************************************************/
34.
35. /********************************************************************************
36. *                                   内部变量
37. *********************************************************************************/
38.
39. /********************************************************************************
40. *                                 内部函数声明
41. *********************************************************************************/
42. static  void  InitSoftware(void);      //初始化 Software
43.
44. static void __stdcall TimeProc(unsigned int uTimerID, unsigned int uMsg, unsigned long dwUser,
45.   unsigned long dw1, unsigned long dw2);   //定时器回调函数
46.
47. static  void  Proc2msTask(void);       //声明一个 2ms 执行一次的函数
48. static  void  Proc1SecTask(void);      //声明一个 1s 执行一次的函数
49.
50. /********************************************************************************
51. *                                 内部函数实现
52. *********************************************************************************/
53. /********************************************************************************
54. * 函数名称: InitSoftware
55. * 函数功能: 所有的软件初始化函数都放在此函数中
56. * 输入参数: void
```

```
57.  * 输出参数: void
58.  * 返 回 值: void
59.  * 创建日期: 2021 年 03 月 01 日
60.  * 注    意:
61.  *********************************************************************************/
62.  static  void  InitSoftware(void)
63.  {
64.    InitRunClock();
65.
66.    //设定初始化时间
67.    SetTimeVal(TIME_VAL_HOUR, 23);
68.    SetTimeVal(TIME_VAL_MIN,  59);
69.    SetTimeVal(TIME_VAL_SEC,  49);
70.  }
71.
72.  /*********************************************************************************
73.  * 函数名称: TimeProc
74.  * 函数功能: 定时器回调函数
75.  * 输入参数: uTimerID, uMsg, dwUser, dw1, dw2, 详见 MSDN
76.  * 输出参数: void
77.  * 返 回 值: void
78.  * 创建日期: 2021 年 03 月 01 日
79.  * 注    意:
80.  *********************************************************************************/
81.  static void __stdcall TimeProc(unsigned int uTimerID, unsigned int uMsg, unsigned long dwUser,
82.    unsigned long dw1, unsigned long dw2)
83.  {
84.    Proc2msTask();
85.  }
86.
87.  /*********************************************************************************
88.  * 函数名称: Proc2msTask
89.  * 函数功能: 处理 2ms 任务
90.  * 输入参数: void
91.  * 输出参数: void
92.  * 返 回 值: void
93.  * 创建日期: 2021 年 03 月 01 日
94.  * 注    意:
95.  *********************************************************************************/
96.  static  void  Proc2msTask(void)
97.  {
98.    static  short s_iCnt500 = 0;
99.
100.   RunClockPer2Ms();
101.
102.   if(s_iCnt500 >=499)
103.   {
104.     Proc1SecTask();
105.     s_iCnt500 = 0;
106.   }
107.   else
108.   {
```

```
109.      s_iCnt500++;
110.    }
111. }
112.
113. /*********************************************************************************
114. * 函数名称: Proc1SecTask
115. * 函数功能: 处理 1s 任务
116. * 输入参数: void
117. * 输出参数: 无
118. * 返 回 值: void
119. * 创建日期: 2021 年 03 月 01 日
120. * 注    意:
121. *********************************************************************************/
122. static  void  Proc1SecTask(void)
123. {
124.   DispTime(GetTimeVal(TIME_VAL_HOUR), GetTimeVal(TIME_VAL_MIN), GetTimeVal(TIME_VAL_SEC));
125. }
126.
127. /*********************************************************************************
128. *                                    API 函数实现
129. *********************************************************************************/
130. /*********************************************************************************
131. * 函数名称: main
132. * 函数功能: 主函数
133. * 输入参数: void
134. * 输出参数: void
135. * 返 回 值: void
136. * 创建日期: 2021 年 03 月 01 日
137. * 注    意:
138. *********************************************************************************/
139. void  main(void)
140. {
141.   InitSoftware();                    //初始化软件
142.
143.   //timeSetEvent 函数说明
144.   //MMRESULT timeSetEvent(
145.   //  UINT            uDelay,          //以 ms 指定事件的周期
146.   //  UINT            uResolution,     //以 ms 指定延时的精度，数值越小定时器事件分辨率越高，
                                            默认值为 1ms
147.   //  LPTIMECALLBACK lpTimeProc,      //指向一个回调函数，即单次事件或周期性事件触发时调用
                                            的函数
148.   //  DWORD_PTR       dwUser,          //存放用户提供的回调数据
149.   //  UINT            fuEvent          //指定定时器事件类型，       TIME_ONESHOT-单次触发，
                                            TIME_PERIODIC-周期性触发
150.   //);
151.
152.   //用户定时器设定，定时器精度为 1ms，每隔 2ms 触发一次定时器，并执行回调函数 TimeProc
153.   timeSetEvent(2, 1, TimeProc, 0, TIME_PERIODIC);
154.
155.   while(1)
156.   {
157.
```

```
158.    }
159. }
```

步骤 5：项目编译和运行

最后，按 F5 键编译并运行程序，在弹出的控制台窗口中，可以看到如图 11-3 所示的运行结果，即每秒打印一个完整的时间值，说明实验成功。

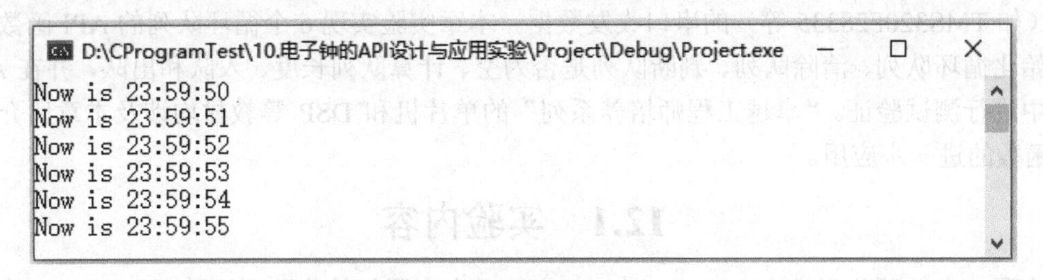

图 11-3　本章项目运行结果

本 章 任 务

2020 年有 366 天，将 2020 年 1 月 1 日作为计数起点，即计数 1，2020 年 12 月 31 日作为计数终点，即计数 366。计数 1 代表“2020 年 1 月 1 日-星期三”，计数 10 代表“2020 年 1 月 10 日-星期五”。参考本章实验，设计一个实验，实现每秒计数递增一次，计数范围为 1～366，并通过 printf 函数每秒打印一次计数对应的年、月、日、星期。例如，初始日期设置为 10，即“2020 年 1 月 10 日-星期五”，则第 1 秒打印“2020 年 1 月 10 日-星期五”、第 2 秒打印“2020 年 1 月 11 日-星期六”，以此类推。

本 章 习 题

1．为什么要使用 DataType.h 文件？

2．对比第 10 章实验和第 11 章实验的异同。

第 12 章　循环队列的 API 设计与应用

在串口通信中常常会用到循环队列，例如，单片机（如 STM32F1 和 STM32F4 等）和 DSP（如 TMS320F28335 等）的串口收发数据。本章实验实现 6 个循环队列的 API 函数，包括初始化循环队列、清除队列、判断队列是否为空、计算队列长度、入队和出队，并在 App.c 文件中进行测试验证。“卓越工程师培养系列”的单片机和 DSP 等教材均涉及本章所介绍的 API 函数的进一步应用。

12.1　实验内容

实现一个循环队列模块，在 Queue.c/.h 文件对中实现初始化队列函数 InitQueue、清除队列函数 ClearQueue、判断队列是否为空函数 QueueEmpty、计算队列长度函数 QueueLength、入队函数 EnQueue 和出队函数 DeQueue，并在 App.c 文件中进行验证。例如，初始化一个包含 30 个元素的队列，先入队 16 个数据，然后出队 25 个数据（实际只能出队 16 个数据），再入队 20 个数据，最后出队 20 个数据。

12.2　实验原理

12.2.1　队列与循环队列

队列是一种先入先出（First In First Out，FIFO）的线性表，它只允许在表的一端插入元素，在另一端取出元素。这与日常生活中的“排队”的概念是一致的，最早进入队列的元素最早离开。在队列中，允许插入的一端称为队尾（rear），允许取出的一端称为队头（front）。

有时为了使用方便，将顺序队列臆造为一个环状的空间，称为循环队列。为了更形象地理解，下面举一个简单的例子。假设指针变量 pQue 指向一个队列，该队列为结构体变量，队列的容量为 8，如图 12-1 所示。循环队列的操作如下：（a）起初，队列为空，队头 pQue→front 和队尾 pQue→rear 均指向地址 0，队列中的元素数量为 0；（b）插入 J0、J1、…、J5 这 6 个元素后，队头 pQue→front 依然指向地址 0，队尾 pQue→rear 指向地址 6，队列中的元素数量为 6；（c）取出 J0、J1、J2、J3 这 4 个元素后，队头 pQue→front 指向地址 4，队尾 pQue→rear 指向地址 6，队列中的元素数量为 2；（d）继续插入 J6、J7、…、J11 这 6 个元素后，队头 pQue→front 指向地址 4，队尾 pQue→rear 也指向地址 4，队列中的元素数量为 8，此时队列为满。

12.2.2　循环队列 Queue 模块函数

Queue 模块共有 6 个 API 函数，分别是 InitQueue、ClearQueue、QueueEmpty、QueueLength、EnQueue 和 DeQueue，下面对这 6 个 API 函数进行介绍。

1．InitQueue

InitQueue 函数的功能是初始化 Queue 模块，具体描述如表 12-1 所示。该函数将 pQue->front、pQue->rear、pQue->elemNum 赋值为 0，将参数 len 赋值给 pQue->bufLen，将参数 pBuf 赋值给 pQue->pBuffer，最后，将指针变量 pQue->pBuffer 指向的元素全部赋初值 0。

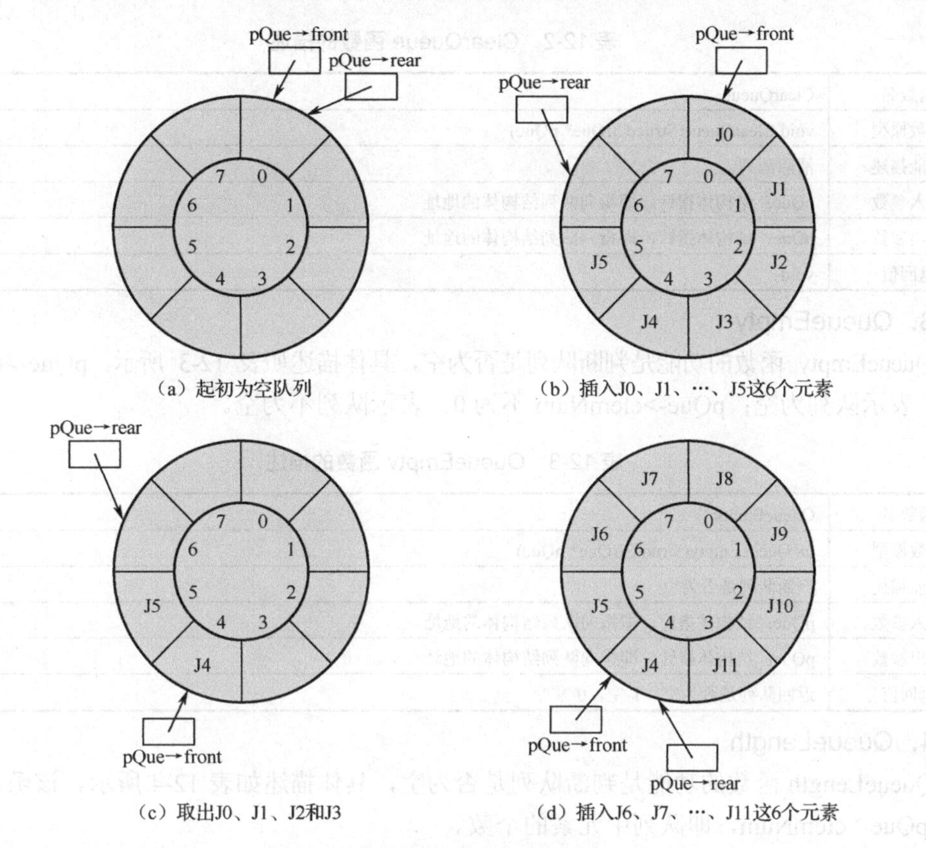

图 12-1　循环队列操作

表 12-1　InitQueue 函数的描述

函数名	InitQueue
函数原型	void InitQueue(StructCirQue* pQue, DATA_TYPE* pBuf, i16 len)
功能描述	初始化 Queue
输入参数	pQue：结构体指针，即指向队列结构体的地址，pBuf-队列的元素存储区地址，len-队列的容量
输出参数	pQue：结构体指针，即指向队列结构体的地址
返回值	void

StructCirQue 结构体定义在 Queue.h 文件中，内容如下：

```
typedef struct
{
  i16       front;          //头指针，队非空时指向队头元素
  i16       rear;           //尾指针，队非空时指向队尾元素的下一个位置
  i16       bufLen;         //队列的总容量
  i16       elemNum;        //当前队列中的元素的数量
  DATA_TYPE *pBuffer;
}StructCirQue;
```

2. ClearQueue

ClearQueue 函数的功能是清除队列，具体描述如表 12-2 所示。该函数将 pQue->front、pQue->rear、pQue->elemNum 赋值为 0。

表 12-2　ClearQueue 函数的描述

函数名	ClearQueue
函数原型	void ClearQueue(StructCirQue* pQue)
功能描述	清除队列
输入参数	pQue：结构体指针，即指向队列结构体的地址
输出参数	pQue：结构体指针，即指向队列结构体的地址
返回值	void

3．QueueEmpty

QueueEmpty 函数的功能是判断队列是否为空，具体描述如表 12-3 所示。pQue->elemNum 为 0，表示队列为空；pQue->elemNum 不为 0，表示队列不为空。

表 12-3　QueueEmpty 函数的描述

函数名	QueueEmpty
函数原型	u8 QueueEmpty(StructCirQue* pQue)
功能描述	判断队列是否为空
输入参数	pQue：结构体指针，即指向队列结构体的地址
输出参数	pQue：结构体指针，即指向队列结构体的地址
返回值	返回队列是否为空，1-空，0-非空

4．QueueLength

QueueLength 函数的功能是判断队列是否为空，具体描述如表 12-4 所示。该函数的返回值为 pQue->elemNum，即队列中元素的个数。

表 12-4　QueueLength 函数的描述

函数名	QueueLength
函数原型	i16 QueueLength(StructCirQue* pQue)
功能描述	判断队列是否为空
输入参数	pQue：结构体指针，即指向队列结构体的地址
输出参数	pQue：结构体指针，即指向队列结构体的地址
返回值	队列中元素的个数

5．EnQueue

EnQueue 函数的功能是插入 len 个元素（存放在起始地址为 pInput 的存储区中）到队列，具体描述如表 12-5 所示。每次插入一个元素，pQue->rear 自增，当 pQue->rear 的值大于或等于数据缓冲区的长度 pQue->bufLen 时，pQue->rear 赋值为 0。注意，当数据缓冲区中的元素数量加上新写入的元素数量超过缓冲区的长度时，缓冲区能接收的元素数量为缓冲区的容量减去缓冲区中已有的元素数量，即 EnQueue 函数对超出的元素采取不理睬的态度。

表 12-5　EnQueue 函数的描述

函数名	EnQueue
函数原型	i16 EnQueue(StructCirQue* pQue, DATA_TYPE* pInput, i16 len)
功能描述	插入 len 个元素（存放在起始地址为 pInput 的存储区中）到队列
输入参数	pQue：结构体指针，即指向队列结构体的地址，pInput-待入队数组的地址，len-期望入队元素的数量
输出参数	pQue：结构体指针，即指向队列结构体的地址
返回值	成功入队的元素的数量

6．DeQueue

DeQueue 函数的功能是从队列中取出 len 个元素，放入起始地址为 pOutput 的存储区，具体描述如表 12-6 所示。每次取出一个元素，pQue->front 自增，当 pQue->front 的值大于或等于数据缓冲区的长度 pQue->bufLen 时，pQue->front 赋值为 0。注意，从队列中提取元素的前提是队列中需要至少有一个元素，当期望取出的元素数量 len 小于或等于队列中元素的数量时，可以按期望取出 len 个元素；否则，只能取出队列中已有的元素。

表 12-6　DeQueue 函数的描述

函数名	DeQueue
函数原型	i16 DeQueue(StructCirQue* pQue, DATA_TYPE* pOutput, i16 len)
功能描述	从队列中取出 len 个元素，放入起始地址为 pOutput 的存储区
输入参数	pQue：结构体指针，即指向队列结构体的地址；pOutput-出队元素存放的数组的地址，len-预期出队元素的数量
输出参数	pQue：结构体指针，即指向队列结构体的地址；pOutput-出队元素存放的数组的地址
返回值	成功出队的元素的数量

12.2.3　for 循环语句

for 语句是比 while 语句更加灵活的循环语句，其一般形式如下：

```
for(表达式 1;表达式 2;表达式 3)
{
    循环体语句;
}
```

表达式 1：循环起始语句，为变量赋初值，只执行一次。可以为零个、一个或多个变量设置初值。

表达式 2：循环条件表达式，每次执行循环体之前先判断表达式 2，若为真，则执行循环体；若为假，则跳出循环。

表达式 3：作为循环的调整，例如，使循环变量递增，它是在执行完循环体后才进行的。

for 循环语句的流程图如图 12-2 所示。

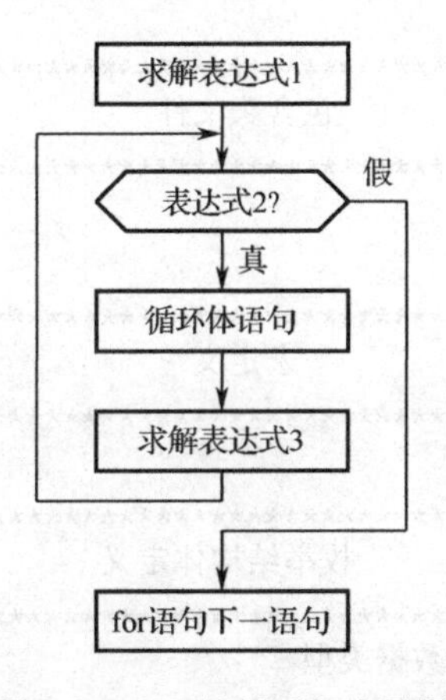

图 12-2　for 循环语句流程图

12.3　实验步骤

步骤 1：复制原始项目

首先，将本书配套资料包的“03.例程资料\Material\11.循环队列的 API 设计与应用实验”文件夹复制到 CProgramTest 文件夹中，然后，双击运行“D:\CProgramTest\11.循环队列的 API 设计与应用实验\Project”文件夹中的 Project.sln 文件。

步骤 2：完善 Queue.h 文件

将程序清单 12-1 中的代码输入 Queue.h 文件中。下面按照顺序对部分语句进行解释。

（1）第 32 行代码：定义一个新的数据类型，这样就可以在代码中用 u8 代替 DATA_TYPE，如果需要更改代码中变量的数据类型（DATA_TYPE），例如，将数据类型由 u8（unsigned char）改为 u16（unsigned short），只需要将“typedef u8 DATA_TYPE”改为“typedef u16 DATA_TYPE”即可。

（2）第 47 至 52 行代码：声明循环队列的 6 个 API 函数。

程序清单 12-1

```
1.  /*********************************************************************************************
2.  * 模块名称: Queue.h
3.  * 摘    要: 队列
4.  * 当前版本: 1.0.0
5.  * 作    者: Leyutek(COPYRIGHT 2018 - 2021 Leyutek. All rights reserved.)
6.  * 完成日期: 2021 年 03 月 01 日
7.  * 内    容:
8.  * 注    意:
9.  *********************************************************************************************
10. * 取代版本:
11. * 作    者:
12. * 完成日期:
13. * 修改内容:
14. * 修改文件:
15. *********************************************************************************************/
16. #ifndef _QUEUE_H_
17. #define _QUEUE_H_
18.
19. /*********************************************************************************************
20. *                                        包含头文件
21. *********************************************************************************************/
22. #include "DataType.h"
23.
24. /*********************************************************************************************
25. *                                          宏定义
26. *********************************************************************************************/
27.
28. /*********************************************************************************************
29. *                                       枚举结构体定义
30. *********************************************************************************************/
31. //定义数据类型，即队列中元素的数据类型
32. typedef u8  DATA_TYPE;
33.
34. //定义循环队列结构体
```

```
35. typedef struct
36. {
37.   i16       front;          //头指针，队非空时指向队头元素
38.   i16       rear;           //尾指针，队非空时指向队尾元素的下一个位置
39.   i16       bufLen;         //队列的总容量
40.   i16       elemNum;        //当前队列中的元素的数量
41.   DATA_TYPE *pBuffer;
42. }StructCirQue;
43.
44. /*********************************************************************************
45. *                                API 函数声明
46. *********************************************************************************/
47. void  InitQueue(StructCirQue* pQue, DATA_TYPE* pBuf, i16 len);   //初始化队列
48. void  ClearQueue(StructCirQue* pQue);                            //清除队列
49. u8    QueueEmpty(StructCirQue* pQue);                 //判断队列是否为空，1-空，0-非空
50. i16   QueueLength(StructCirQue* pQue);                //返回队列中元素个数，即为队列的长度
51. i16   EnQueue(StructCirQue* pQue, DATA_TYPE* pInput, i16 len);   //入队 len 个元素
52. i16   DeQueue(StructCirQue* pQue, DATA_TYPE* pOutput, i16 len);  //出队 len 个元素
53.
54. #endif
```

步骤 3：完善 Queue.c 文件

将程序清单 12-2 中的代码输入 Queue.c 文件中。该文件的主要功能是定义循环队列的 6 个 API 函数，这些函数已经在 12.2.2 节进行了解释，同时代码中有详细的注释，这里不再介绍。

程序清单 12-2

```
1.  /*********************************************************************************
2.  * 模块名称：Queue.c
3.  * 摘    要：队列
4.  * 当前版本：1.0.0
5.  * 作    者：Leyutek(COPYRIGHT 2018 - 2021 Leyutek. All rights reserved.)
6.  * 完成日期：2021 年 03 月 01 日
7.  * 内    容：
8.  * 注    意：
9.  *********************************************************************************
10. * 取代版本：
11. * 作    者：
12. * 完成日期：
13. * 修改内容：
14. * 修改文件：
15. *********************************************************************************/
16.
17. /*********************************************************************************
18. *                                包含头文件
19. *********************************************************************************/
20. #include "Queue.h"
21.
22. /*********************************************************************************
23. *                                  宏定义
24. *********************************************************************************/
```

```
25.
26. /*****************************************************************************************
27. *                                    枚举结构体定义
28. *****************************************************************************************/
29.
30. /*****************************************************************************************
31. *                                    内部变量
32. *****************************************************************************************/
33.
34. /*****************************************************************************************
35. *                                    内部函数声明
36. *****************************************************************************************/
37.
38. /*****************************************************************************************
39. *                                    内部函数实现
40. *****************************************************************************************/
41.
42. /*****************************************************************************************
43. *                                    API 函数实现
44. *****************************************************************************************/
45.
46. /*****************************************************************************************
47. * 函数名称: InitQueue
48. * 函数功能: 初始化队列
49. * 输入参数: pQue-结构体指针，即指向队列结构体的地址，pBuf-队列的元素存储区地址，len-队列
            的容量
50. * 输出参数: pQue-结构体指针，即指向队列结构体的地址
51. * 返 回 值: void
52. * 创建日期: 2021 年 03 月 01 日
53. * 注    意:
54. *****************************************************************************************/
55. void  InitQueue(StructCirQue* pQue, DATA_TYPE* pBuf, i16 len)
56. {
57.   i16 i;
58.
59.   pQue->front          = 0;      //队头
60.   pQue->rear           = 0;      //队尾
61.   pQue->bufLen         = len;    //buffer 长度（容量）
62.   pQue->elemNum        = 0;      //当前队列中元素的个数
63.   pQue->pBuffer        = pBuf;   //指针变量 pBuf 赋给指针变量 pQue ->pBuffer
64.
65.   for(i = 0; i < len; i++)
66.   {
67.     pQue->pBuffer[i]  = 0;       //对指针变量 pQue->pBuffer 所指向的元素赋初值 0
68.   }
69. }
70.
71. /*****************************************************************************************
```

```
72. * 函数名称: ClearQueue
73. * 函数功能: 清除队列
74. * 输入参数: pQue-结构体指针，即指向队列结构体的地址
75. * 输出参数: pQue-结构体指针，即指向队列结构体的地址
76. * 返 回 值: void
77. * 创建日期: 2021 年 03 月 01 日
78. * 注    意:
79. *******************************************************************************/
80. void  ClearQueue(StructCirQue* pQue)
81. {
82.   pQue->front         = 0;       //队头
83.   pQue->rear          = 0;       //队尾
84.   pQue->elemNum       = 0;       //当前的数据长度
85. }
86.
87. /*******************************************************************************
88. * 函数名称: QueueEmpty
89. * 函数功能: 判断队列是否为空，1-空，0-非空
90. * 输入参数: pQue-结构体指针，即指向队列结构体的地址
91. * 输出参数: pQue-结构体指针，即指向队列结构体的地址
92. * 返 回 值: 返回队列是否为空，1-空，0-非空
93. * 创建日期: 2021 年 03 月 01 日
94. * 注    意:
95. *******************************************************************************/
96. u8    QueueEmpty(StructCirQue* pQue)
97. {
98.   return(0 == pQue->elemNum);
99. }
100.
101. /*******************************************************************************
102. * 函数名称: QueueLength
103. * 函数功能: 返回队列中元素个数，即为队列的长度
104. * 输入参数: pQue-结构体指针，即指向队列结构体的地址
105. * 输出参数: pQue-结构体指针，即指向队列结构体的地址
106. * 返 回 值: 队列中元素的个数
107. * 创建日期: 2021 年 03 月 01 日
108. * 注    意:
109. *******************************************************************************/
110. i16   QueueLength(StructCirQue* pQue)
111. {
112.   return(pQue->elemNum);
113. }
114.
115. /*******************************************************************************
116. * 函数名称: EnQueue
117. * 函数功能: 插入 len 个元素（存放在起始地址为 pInput 的存储区中）到队列
118. * 输入参数: pQue-结构体指针，即指向队列结构体的地址，pInput-待入队数组的地址，len-期望
            入队元素的数量
```

```
119. * 输出参数: pQue-结构体指针，即指向队列结构体的地址
120. * 返 回 值: 成功入队的元素的数量
121. * 创建日期: 2021 年 03 月 01 日
122. * 注    意: 每插入一个元素，rear 自增，当 rear 的值大于或等于数据 buffer 的长度 bufLen 时，rear 值
123. * 变为零。注意，当数据 buffer 中的元素数量加上新写入的元素数量超过 buffer 长度时，buffer 能接收的
124. * 元素数量为 buffer 的容量-buffer 中已有的元素数量，即 EnQueue 函数对于超出的元素采取不理睬的态度。
125. ***********************************************************************************/
126. i16 EnQueue(StructCirQue* pQue, DATA_TYPE* pInput, i16 len)
127. {
128.   i16 wLen = 0;                   //待入队的元素有 len 个，wLen 从 0 增加到 len-1
129.
130.   while((pQue->elemNum < pQue->bufLen) && (wLen < len))
131.   {
132.     pQue->pBuffer[pQue->rear] = pInput[wLen]; //将待入队的第 wLen 个元素 pInput[wLen]插入队列
133.     pQue->rear++;                 //队尾 rear 自增，即指向队尾元素的下一个位置
134.
135.     if(pQue->rear >= pQue->bufLen)
136.     {
137.       pQue->rear = 0;             //如果队尾元素的下一个位置为 pQue->bufLen，则 rear 指向队头，
                                        循环体现在此处
138.     }
139.
140.     wLen++;
141.     pQue->elemNum++;              //当前队列中的元素总数
142.   }
143.
144.   return wLen;                    //返回值 wLen 为 0，表示没有元素入队
145. }
146.
147. /**********************************************************************************
148. * 函数名称: DeQueue
149. * 函数功能: 从队列中取出 len 个元素，放入起始地址为 pOutput 的存储区
150. * 输入参数: pQue-结构体指针，即指向队列结构体的地址，len-预期出队元素的数量
151. * 输出参数: pQue-结构体指针，即指向队列结构体的地址，pOutput-出队元素存放的数组的地址
152. * 返 回 值: 成功出队的元素的数量
153. * 创建日期: 2021 年 03 月 01 日
154. * 注    意: 每次取出一个元素，front 自增，当 front 的值大于或等于数据 buffer 的长度 bufLen 时，front
155. * 值变为零。注意，从队列中提取元素的前提是队列中至少要有一个元素，当期望取出的元素数量 len 小于
156. * 或等于队列中元素的数量时，可以按期望取出 len 个元素，否则，只能取出队列中已有的所有元素。
157. ***********************************************************************************/
158. i16 DeQueue(StructCirQue* pQue, DATA_TYPE* pOutput, i16 len)
159. {
160.   i16 rLen = 0;                   //期望取出 len 个元素，最终能取出 rLen 个元素
161.
162.   while((pQue->elemNum > 0) && (rLen < len))
163.   {
164.     pOutput[rLen] = pQue->pBuffer[pQue->front];
165.     pQue->front++;
```

```
166.
167.     if( pQue->front >= pQue->bufLen )
168.     {
169.       pQue->front = 0;            //如果队头元素的下一个位置为 pQue->bufLen，则 front 指向
                                         队头，循环体现在此处
170.     }
171.
172.     rLen++;
173.     pQue->elemNum--;              //当前队列中的元素总数
174.   }
175.
176.   return rLen;                    //如果返回值 rLen 为 0，表示队列中没有元素
177. }
```

步骤 4：完善 App.c 文件

将程序清单 12-3 中的代码输入 App.c 文件中。下面按照顺序对部分语句进行解释。

（1）第 59 至 79 行代码：定义一个循环队列 cirQueForTest，数组 bufForTest 用于保存该队列中的元素，该队列可容纳 30 个元素。在测试该队列的过程中，通过数组 arrBufForInput 向该队列中插入元素，出队元素保存到数组 arrBufForOutput 中。InitQueue 函数用于对该队列进行初始化。

（2）第 82 至 90 行代码：通过 EnQueue 函数向循环队列 cirQueForTest 中入队 16 个数据（保存在数组 arrBufForInput 中），并打印出该循环队列中的所有元素。

（3）第 92 至 101 行代码：通过 DeQueue 函数尝试从循环队列 cirQueForTest 中出队 25 个数据到数组 arrBufForOutput 中，并打印成功出队的数量和数据。

（4）第 103 至 114 行代码：通过 EnQueue 函数向循环队列 cirQueForTest 中入队 20 个数据（保存在数组 arrBufForInput 中），并打印出该循环队列中的所有元素。

（5）第 116 至 126 行代码：通过 DeQueue 函数尝试从循环队列 cirQueForTest 中出队 20 个数据到数组 arrBufForOutput 中，并打印成功出队的数量和数据。

程序清单 12-3

```
1.  /*********************************************************************************************
2.  * 模块名称: App.c
3.  * 摘    要: 测试队列
4.  * 当前版本: 1.0.0
5.  * 作    者: Leyutek(COPYRIGHT 2018 - 2021 Leyutek. All rights reserved.)
6.  * 完成日期: 2021 年 03 月 01 日
7.  * 内    容:
8.  * 注    意:
9.  *********************************************************************************************
10. * 取代版本:
11. * 作    者:
12. * 完成日期:
13. * 修改内容:
14. * 修改文件:
15. *********************************************************************************************/
16.
17. /*********************************************************************************************
```

```
18. *                                  包含头文件
19. *********************************************************************************/
20. #include <stdio.h>
21. #include <stdlib.h>
22. #include "Queue.h"
23.
24. /*********************************************************************************
25. *                                    宏定义
26. *********************************************************************************/
27.
28. /*********************************************************************************
29. *                                枚举结构体定义
30. *********************************************************************************/
31.
32. /*********************************************************************************
33. *                                   内部变量
34. *********************************************************************************/
35.
36. /*********************************************************************************
37. *                                 内部函数声明
38. *********************************************************************************/
39.
40. /*********************************************************************************
41. *                                 内部函数实现
42. *********************************************************************************/
43.
44. /*********************************************************************************
45. *                                 API 函数实现
46. *********************************************************************************/
47.
48. /*********************************************************************************
49. * 函数名称: main
50. * 函数功能: 主函数
51. * 输入参数: void
52. * 输出参数: void
53. * 返 回 值: void
54. * 创建日期: 2021 年 03 月 01 日
55. * 注    意: 该函数的主要目的是测试循环队列是否正确
56. *********************************************************************************/
57. void main(void)
58. {
59.   StructCirQue    cirQueForTest;          //循环队列
60.   DATA_TYPE i;
61.   DATA_TYPE bufForTest[30];               //循环队列中的 buffer，可容纳 30 个元素
62.
63.   i16       len;                          //长度
64.
65.   DATA_TYPE arrBufForOutput[30];          //从队列取出的数据放入此数组中
66.   DATA_TYPE arrBufForInput[60];           //入队的数据先放到此处
67.   DATA_TYPE dataForTest;                  //便于 printf 调用
```

```
68.
69.    for(i = 0; i < 60; i++)
70.    {
71.      arrBufForInput[i] = 60 + i * 2 + 1;
72.    }
73.
74.    for(i = 0; i < 30; i++)
75.    {
76.      arrBufForOutput[i] = 0;
77.    }
78.
79.    InitQueue(&cirQueForTest, bufForTest, 30);
80.
81.
82.    //入队 16 个数据
83.    printf("**************TEST1-入队 16 个数据******************\n");
84.    EnQueue(&cirQueForTest, arrBufForInput, 16);
85.
86.    for(i = 0; i < 30; i++)
87.    {
88.      dataForTest = cirQueForTest.pBuffer[i];
89.      printf("addr=%d, dataInput=%d\n", i, dataForTest);
90.    }
91.
92.    //出队 25 个数据，实际只能出队 16 个数据
93.    printf("**************TEST2-出队 25 个数据，实际只能出队 16 个数据******************\n");
94.    len = DeQueue(&cirQueForTest, arrBufForOutput, 25);
95.
96.    printf("Num of DeQueue: %d\n", len);
97.
98.    for(i = 0; i < len; i++)
99.    {
100.     printf("i=%d, dataOuput=%d\n", i, arrBufForOutput[i]);
101.   }
102.
103.   //入队 20 个数据
104.   printf("**************TEST3-入队 20 个数据******************\n");
105.
106.   len = EnQueue(&cirQueForTest, &arrBufForInput[20], 20);
107.
108.   printf("Num of EnQueue: %d\n", len);
109.
110.   for(i = 0; i < 30; i++)
111.   {
112.     dataForTest = cirQueForTest.pBuffer[i];
113.     printf("addr=%d, dataInput=%d\n", i, dataForTest);
114.   }
115.
116.   //出队 20 个数据
117.   printf("**************TEST4-出队 20 个数据******************\n");
```

```
118.
119.   len = DeQueue(&cirQueForTest, arrBufForOutput, 20);
120.
121.   printf("Num of DeQueue: %d\n", len);
122.
123.   for(i = 0; i < len; i++)
124.   {
125.     printf("i=%d, dataOuput=%d\n", i, arrBufForOutput[i]);
126.   }
127.
128.   system("pause");
129. }
```

步骤 5：项目编译和运行

最后，按 F5 键编译并运行程序，在弹出的控制台窗口中，可以看到如图 12-3 所示的运行结果，分析如下：①入队 16 个数据（61、63、…、91）到循环队列，这样该循环队列中就有 16 个有效数据，其余 14 个存储空间为默认值 0；②尝试出队 25 个数据，由于该队列中只有 16 个有效数据，因此只能出队 16 个数据，并且按照先进先出的顺序。

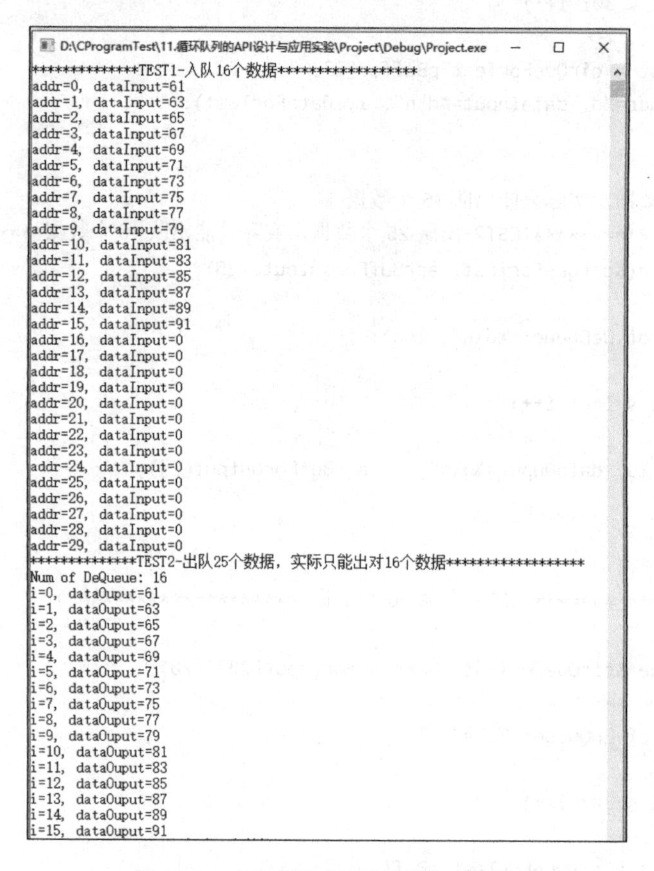

图 12-3　本章项目运行结果第一部分

更进一步分析：①入队 20 个数据（101、103、…、139）到循环队列，这样该循环队列中就有 20 个有效数据；②出队 20 个数据，这样这些数据就全部出队，同样按照先进先出的顺序，如图 12-4 所示，说明实验成功。

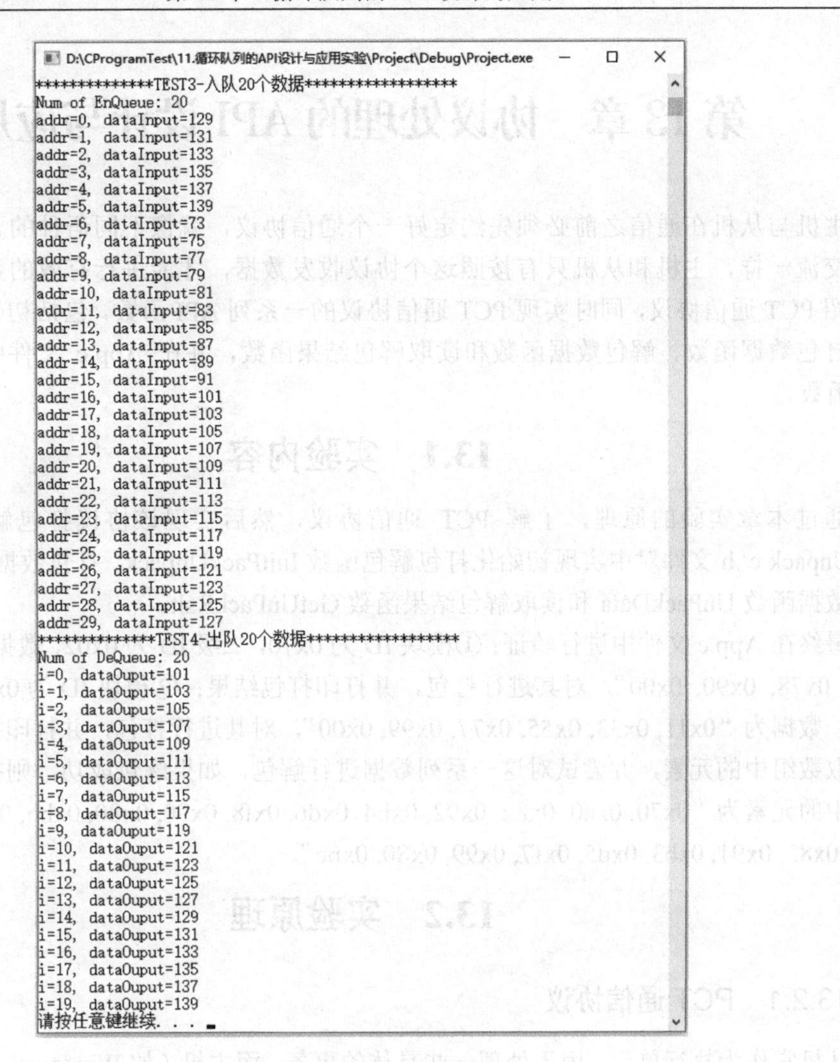

图 12-4　本章项目运行结果第二部分

本 章 任 务

基于本章实验的 Queue 模块，初始化一个队列，该队列可容纳 20 个元素，且元素的类型为 short 型，先入队 10 个元素（依次为 1～10），再入队 5 个元素（1、5、10、20、25），然后出队 12 个元素，并将出队的这 12 个元素依次通过 printf 函数打印出来，同时，计算队列中元素的个数。

本 章 习 题

1．什么是队列？什么是循环队列？

2．Queue 模块中都有哪些 API 函数？简述每个 API 函数的功能。

第13章　协议处理的API设计与应用

主机与从机在通信之前必须先约定好一个通信协议，就像不同语种的人约定好用某种语言交流一样，主机和从机只有按照这个协议收发数据，才能进行有效的数据传输。本章将介绍PCT通信协议，同时实现PCT通信协议的一系列API函数，包括初始化打包解包函数、打包数据函数、解包数据函数和读取解包结果函数，并在App.c文件中测试验证这些API函数。

13.1　实验内容

通过本章实验的原理，了解PCT通信协议，然后实现该协议打包解包模块，即在PackUnpack.c/.h文件对中实现初始化打包解包函数InitPackUnpack、打包数据函数PackData、解包数据函数UnPackData和读取解包结果函数GetUnPackRslt。

最终在App.c文件中进行验证：①模块ID为0x70，二级ID为0x02，数据为“0x12, 0x34, 0x56, 0x78, 0x90, 0x00”，对其进行打包，并打印打包结果；②模块ID为0x71，二级ID为0x02，数据为“0x11, 0x33, 0x55, 0x77, 0x99, 0x00”，对其进行打包，并打印打包结果；③依次读取数组中的元素，并尝试对这一系列数据进行解包，如果解包成功，则打印解包结果，数组中的元素为“0x70, 0xa0, 0x82, 0x92, 0xb4, 0xd6, 0xf8, 0x90, 0x80, 0xb6, 0x22, 0x41, 0x71, 0xa0, 0x82, 0x91, 0xb3, 0xd5, 0xf7, 0x99, 0x80, 0xbc”。

13.2　实验原理

13.2.1　PCT通信协议

从机常作为执行单元，用于处理一些具体的事务，而主机（如Windows、Linux、Android和emWin平台等）常用于与从机进行交互，向从机发送命令，或处理来自从机的数据，如图13-1所示。

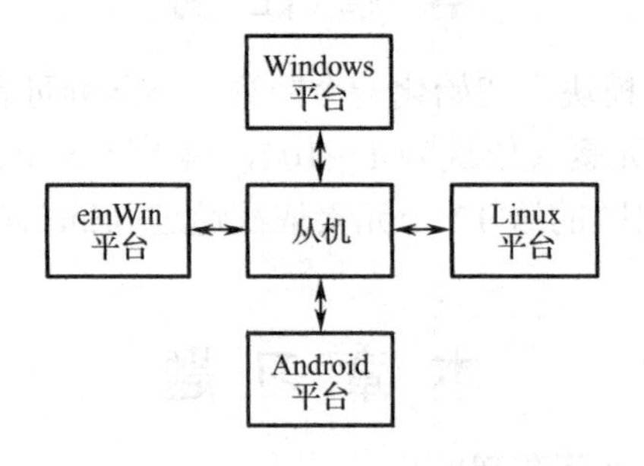

图13-1　主机与从机交互框图

主机与从机之间的通信过程如图13-2所示。主机向从机发送命令的具体过程是：①主机对待发命令进行打包；②主机通过通信设备（串口、蓝牙、Wi-Fi等）将打包好的命令发送出去；③从机在接收到命令之后，对命令进行解包；④从机按照相应的命令执行任务。

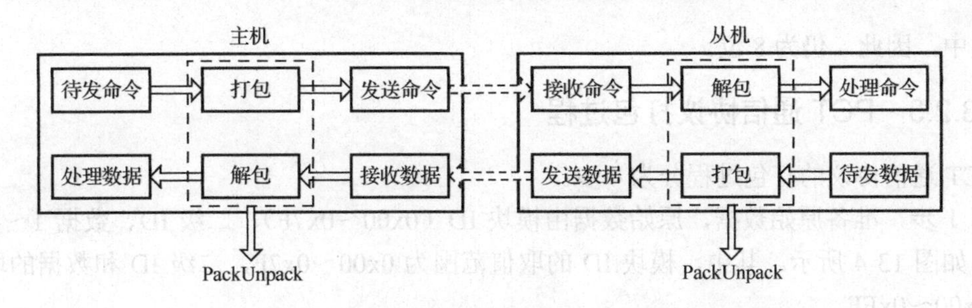

图 13-2　主机与从机之间的通信过程（打包/解包框架图）

从机向主机发送数据的具体过程是：①从机对待发数据进行打包；②从机通过通信设备（串口、蓝牙、Wi-Fi 等）将打包好的数据发送出去；③主机在接收到数据之后，对数据进行解包；④主机对接收到的数据进行处理，如进行计算、显示等。

13.2.2　PCT 通信协议格式

在主机与从机的通信过程中，主机和从机有一个共同的模块，即打包解包模块（PackUnpack），该模块遵循某种通信协议。通信协议有很多种，本章实验采用的 PCT 通信协议由本书作者设计，该协议可由 C、C++、C#、Java 等编程语言实现。打包后的 PCT 通信协议的数据包格式如图 13-3 所示。

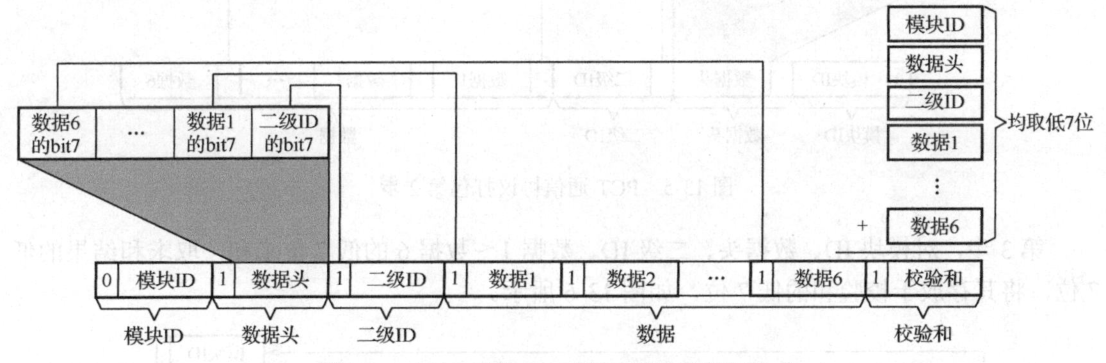

图 13-3　打包后的 PCT 通信协议的数据包格式

PCT 通信协议规定：

（1）数据包由 1 字节模块 ID+1 字节数据头+1 字节二级 ID+6 字节数据+1 字节校验和构成，共计 10 字节。

（2）数据包中有 6 个数据，每个数据为 1 字节。

（3）模块 ID 的最高位 bit7 固定为 0。

（4）模块 ID 的取值范围为 0x00～0x7F，最多有 128 种类型。

（5）数据头的最高位 bit7 固定为 1，数据头的低 7 位按照从低位到高位的顺序，依次存放二级 ID 的最高位 bit7、数据 1 的最高位 bit7、数据 2 的最高位 bit7、数据 3 的最高位 bit7、数据 4 的最高位 bit7、数据 5 的最高位 bit7 和数据 6 的最高位 bit7。

（6）校验和的低 7 位为模块 ID+数据头+二级 ID+数据 1+数据 2+…+数据 6 求和的结果（取低 7 位）。

（7）二级 ID、数据 1～数据 6 和校验和的最高位 bit7 固定为 1。注意，并不是说二级 ID、数据 1～数据 6 和校验和只有 7 位，而是在打包后，它们的低 7 位位置不变，最高位均位于数据头中，因此，仍为 8 位。

13.2.3　PCT 通信协议打包过程

PCT 通信协议的打包过程分为 4 步。

第 1 步，准备原始数据，原始数据由模块 ID（0x00～0x7F）、二级 ID、数据 1～数据 6 组成，如图 13-4 所示。其中，模块 ID 的取值范围为 0x00～0x7F，二级 ID 和数据的取值范围为 0x00～0xFF。

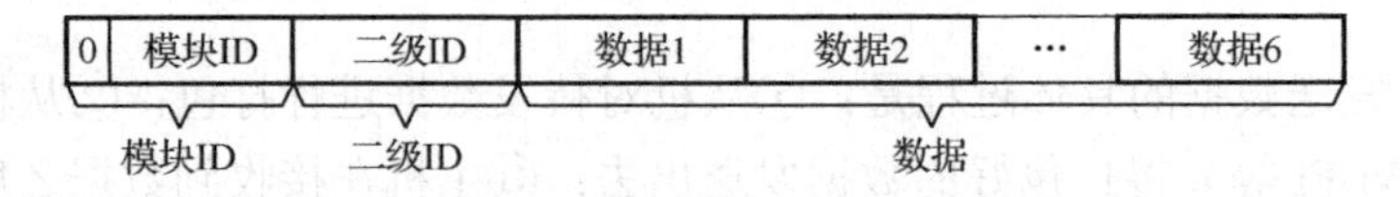

图 13-4　PCT 通信协议打包第 1 步

第 2 步，依次取出二级 ID、数据 1～数据 6 的最高位 bit7，将其存放于数据头的低 7 位，按照从低位到高位的顺序依次存放二级 ID、数据 1～数据 6 的最高位 bit7，如图 13-5 所示。

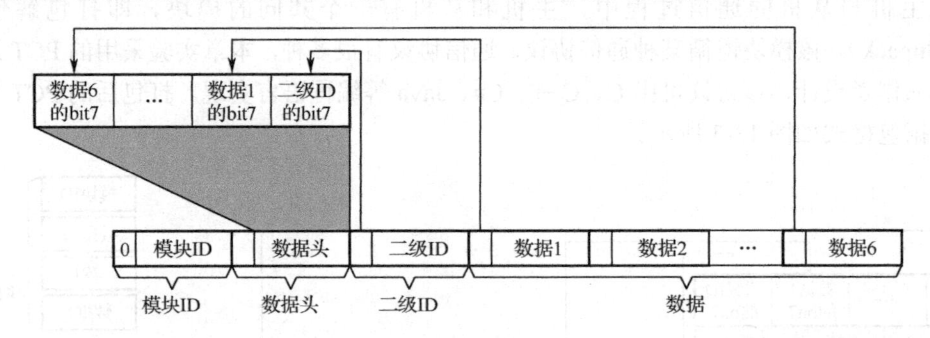

图 13-5　PCT 通信协议打包第 2 步

第 3 步，对模块 ID、数据头、二级 ID、数据 1～数据 6 的低 7 位求和，取求和结果的低 7 位，将其存放于校验和的低 7 位，如图 13-6 所示。

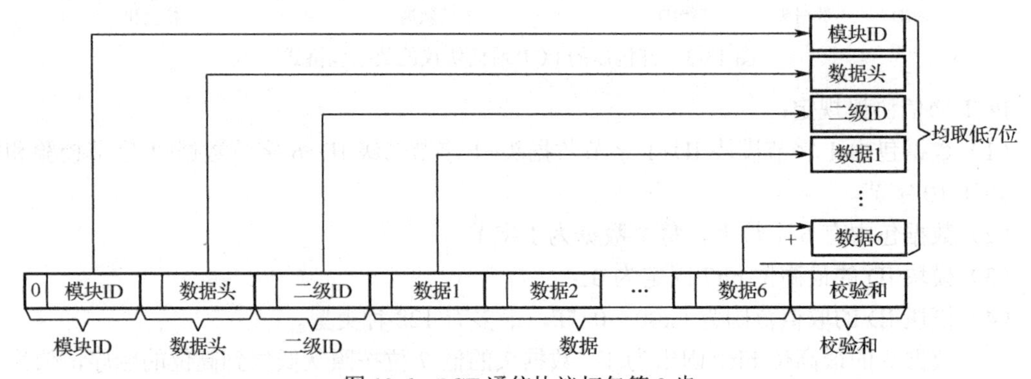

图 13-6　PCT 通信协议打包第 3 步

第 4 步，将数据头、二级 ID、数据 1～数据 6 和校验和的最高位置 1，如图 13-7 所示。

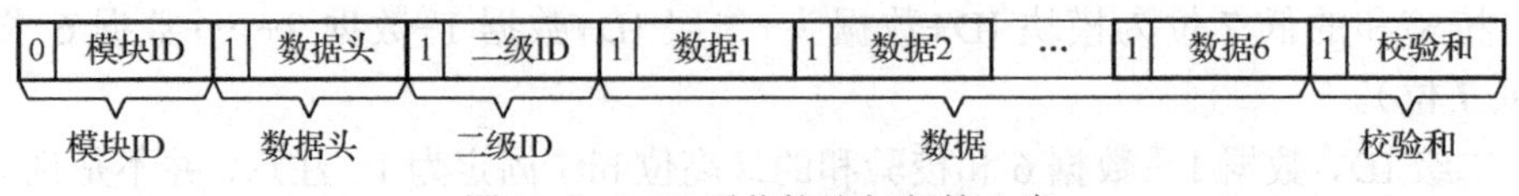

图 13-7　PCT 通信协议打包第 4 步

13.2.4　PCT 通信协议解包过程

PCT 通信协议的解包过程也分为 4 步。

第 1 步，准备解包前的数据包，原始数据包由模块 ID、数据头、二级 ID、数据 1～数据 6 及校验和组成，如图 13-8 所示。其中，模块 ID 的最高位为 0，其余字节的最高位均为 1。

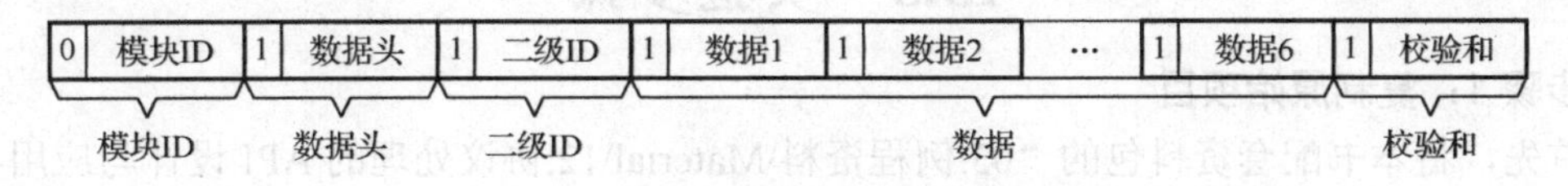

图 13-8　PCT 通信协议解包第 1 步

第 2 步，对模块 ID、数据头、二级 ID、数据 1～数据 6 的低 7 位求和，如图 13-9 所示，取求和结果的低 7 位与数据包的校验和低 7 位进行对比，如果两个值相等，则说明校验正确。

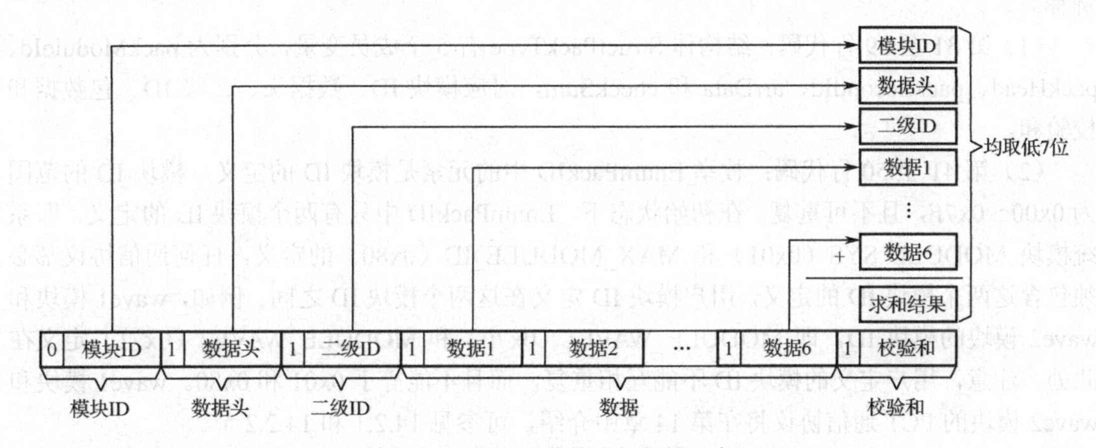

图 13-9　PCT 通信协议解包第 2 步

第 3 步，数据头的最低位 bit0 与二级 ID 的低 7 位拼接之后作为最终的二级 ID，数据头的 bit1 与数据 1 的低 7 位拼接之后作为最终的数据 1，数据头的 bit2 与数据 2 的低 7 位拼接之后作为最终的数据 2，以此类推，如图 13-10 所示。

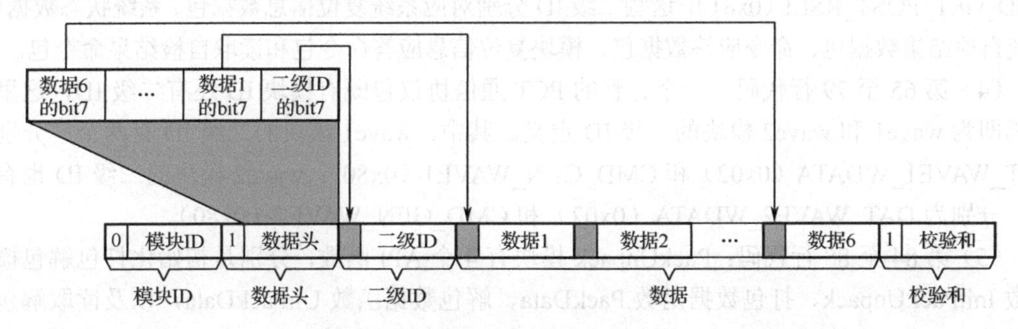

图 13-10　PCT 通信协议解包第 3 步

第 4 步，图 13-11 所示即为解包后的结果，由模块 ID、二级 ID、数据 1～数据 6 组成。其中，模块 ID 的取值范围为 0x00～0x7F，二级 ID 和数据的取值范围为 0x00～0xFF。

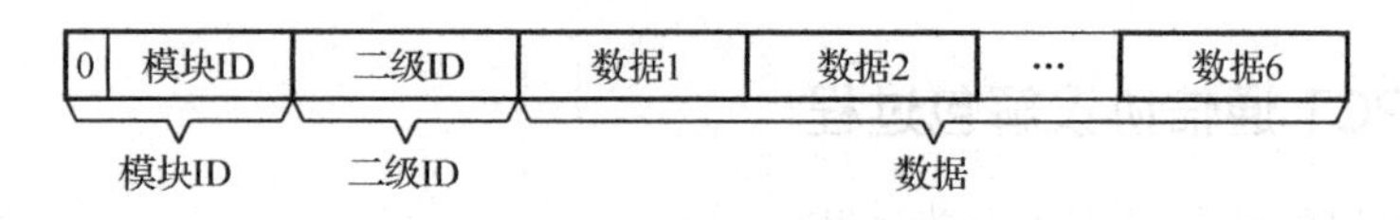

图 13-11　PCT 通信协议解包第 4 步

13.3　实验步骤

步骤 1：复制原始项目

首先，将本书配套资料包的“03.例程资料\Material\12.协议处理的 API 设计与应用实验”文件夹复制到 CProgramTest 文件夹中，然后，双击运行“D:\CProgramTest\12.协议处理的 API 设计与应用实验\Project”文件夹中的 Project.sln 文件。

步骤 2：完善 PackUnpack.h 文件

将程序清单 13-1 中的代码输入 PackUnpack.h 文件中。下面按照顺序对部分语句进行解释。

（1）第 31 至 39 行代码：结构体 StructPackType 有 5 个成员变量，分别为 packModuleId、packHead、packSecondId、arrData 和 checkSum，对应模块 ID、数据头、二级 ID、包数据和校验和。

（2）第 41 至 50 行代码：枚举 EnumPackID 中的元素是模块 ID 的定义，模块 ID 的范围为 0x00～0x7F，且不可重复。在初始状态下，EnumPackID 中只有两个模块 ID 的定义，即系统模块 MODULE_SYS（0x01）和 MAX_MODULE_ID（0x80）的定义，任何通信协议都必须包含这两个模块 ID 的定义，用户模块 ID 定义在这两个模块 ID 之间。例如，wave1 模块和 wave2 模块的模块 ID，即 MODULE_WAVE1（0x70）和 MODULE_WAVE2（0x71）定义在此处。注意，用户定义的模块 ID 不能互相重复，而且不能等于 0x01 和 0x80。wave1 模块和 wave2 模块的 PCT 通信协议将在第 14 章中介绍，可参见 14.2.1 和 14.2.2 节。

（3）第 52 至 63 行代码：枚举 EnumSysSecondID 中的元素是系统模块二级 ID 的定义，系统模块二级 ID 的范围为 0x00～0xFF，不同模块的二级 ID 可以重复。在初始状态下，EnumSysSecondID 中只有 6 个二级 ID 的定义，即 DAT_RST（0x01）、DAT_SYS_STS（0x02）、DAT_SELF_CHECK（0x03）、DAT_CMD_ACK（0x04）、CMD_RST_ACK（0x80）和 CMD_GET_POST_RSLT（0x81），这些二级 ID 分别对应系统复位信息数据包、系统状态数据包、系统自检结果数据包、命令应答数据包、模块复位信息应答命令包和读取自检结果命令包。

（4）第 65 至 79 行代码：一个完整的 PCT 通信协议包既有模块 ID 还有二级 ID，这里的代码即为 wave1 和 wave2 模块的二级 ID 定义。其中，wave1 模块的二级 ID 有两个，分别为 DAT_WAVE1_WDATA（0x02）和 CMD_GEN_WAVE1（0x80）；wave2 模块的二级 ID 也有两个，分别为 DAT_WAVE2_WDATA（0x02）和 CMD_GEN_WAVE2（0x80）。

（5）第 84 至 88 行代码：PackUnpack 模块有 4 个 API 函数，分别是初始化打包解包模块函数 InitPackUnpack、打包数据函数 PackData、解包数据函数 UnPackData，以及读取解包结果函数 GetUnPackRslt。

程序清单 13-1

```
1.  /*********************************************************************************************
2.  * 模块名称：PackUnpack.h
3.  * 摘    要：PCT 协议的打包解包模块
```

```
4.  * 当前版本: 1.0.0
5.  * 作    者: Leyutek(COPYRIGHT 2018 - 2021 Leyutek. All rights reserved.)
6.  * 完成日期: 2021 年 03 月 01 日
7.  * 内    容:
8.  * 注    意:
9.  ********************************************************************************
10. * 取代版本:
11. * 作    者:
12. * 完成日期:
13. * 修改内容:
14. * 修改文件:
15. *******************************************************************************/
16. #ifndef _PACK_UNPACK_H_
17. #define _PACK_UNPACK_H_
18.
19. /*******************************************************************************
20. *                                  包含头文件
21. *******************************************************************************/
22. #include "DataType.h"
23.
24. /*******************************************************************************
25. *                                    宏定义
26. *******************************************************************************/
27.
28. /*******************************************************************************
29. *                                枚举结构体定义
30. *******************************************************************************/
31. //包类型结构体
32. typedef struct
33. {
34.   u8 packModuleId;                      //模块 ID
35.   u8 packHead;                          //数据头
36.   u8 packSecondId;                      //二级 ID
37.   u8 arrData[6];                        //包数据
38.   u8 checkSum;                          //校验和
39. }StructPackType;
40.
41. //枚举定义，定义模块 ID，0x00-0x7F，不可以重复
42. typedef enum
43. {
44.   MODULE_SYS      = 0x01,               //系统信息
45.
46.   MODULE_WAVE1    = 0x70,               //波形 1 信息
47.   MODULE_WAVE2    = 0x71,               //波形 2 信息
48.
49.   MAX_MODULE_ID   = 0x80
50. }EnumPackID;
51.
52. //定义二级 ID，0x00-0xFF，因为是分属于不同的模块 ID，因此不同模块 ID 的二级 ID 可以重复
53. //系统模块的二级 ID
54. typedef enum
55. {
```

```
56.   DAT_RST          = 0x01,              //系统复位信息
57.   DAT_SYS_STS      = 0x02,              //系统状态
58.   DAT_SELF_CHECK   = 0x03,              //系统自检结果
59.   DAT_CMD_ACK      = 0x04,              //命令应答
60.   CMD_RST_ACK      = 0x80,              //模块复位信息应答
61.   CMD_GET_POST_RSLT = 0x81,             //读取自检结果
62.   //用户二级 ID
63. }EnumSysSecondID;
64.
65. //wave1 模块的二级 ID
66. typedef enum
67. {
68.   DAT_WAVE1_WDATA = 0x02,               //波形 1 的波形数据
69.
70.   CMD_GEN_WAVE1   = 0x80,               //生成波形 1 命令
71. }EnumWave1SecondID;
72.
73. //wave2 模块的二级 ID
74. typedef enum
75. {
76.   DAT_WAVE2_WDATA = 0x02,               //波形 2 的波形数据
77.
78.   CMD_GEN_WAVE2   = 0x80,               //生成波形 2 命令
79. }EnumWave2SecondID;
80.
81. /*********************************************************************************************
82. *                                       API 函数声明
83. *********************************************************************************************/
84. void  InitPackUnpack(void);             //初始化打包解包模块
85. u8    PackData(StructPackType* pPT); //对数据进行打包，1-打包成功，0-打包失败
86. u8    UnPackData(u8 data);              //对数据进行解包，1-解包成功，0-解包失败
87.
88. StructPackType  GetUnPackRslt(void); //读取解包后数据包
89.
90. #endif
```

步骤 3：完善 PackUnpack.c 文件

将程序清单 13-2 中的代码输入 PackUnpack.c 文件中。下面按照顺序对部分语句进行解释。

（1）第 55 至 97 行代码：PackWithCheckSum 函数的执行过程可参见 13.2.3 节中的 PCT 通信协议打包过程。

（2）第 99 至 136 行代码：UnpackWithCheckSum 函数的执行过程参见 13.2.4 节中的 PCT 通信协议解包过程。

（3）第 142 至 169 行代码：InitPackUnpack 函数用于对内部静态变量赋初值。

（4）第 171 至 191 行代码：在 PackData 函数中，待打包的数据首地址为 pPT，只有待打包数据包的模块 ID 在 0x00～0x7F 之间（包含 0x00 和 0x7F），才通过调用 PackWithCheckSum 函数对待打包数据进行打包处理。

（5）第 193 至 239 行代码：在 UnPackData 函数中，当接收到完整的首字节为有效模块 ID（0x00～0x7F）的 10 字节数据时，进一步调用 UnpackWithCheckSum 函数对这 10 字节数据进行解包处理。UnPackData 函数的返回值为 1，表示解析到一个有效包，这时，就可以通

过调用 GetUnPackRslt 函数将解包后的数据取走。

（6）第 241 至 253 行代码：GetUnPackResult 函数用于获取解包结果，解包结果由模块 ID、二级 ID 和 6 字节数据组成，它们保存在结构体变量 s_ptPack 中，该函数的返回值即为解包结果。

程序清单 13-2

```
1.  /*********************************************************************************************
2.  * 模块名称: PackUnpack.c
3.  * 摘    要: PCT 协议的打包解包模块
4.  * 当前版本: 1.0.0
5.  * 作    者: Leyutek(COPYRIGHT 2018 - 2021 Leyutek. All rights reserved.)
6.  * 完成日期: 2021 年 03 月 01 日
7.  * 内    容:
8.  *              (1) 数据包格式：模块 ID+数据头+二级 ID+数据 1、…、数据 6+校验和，数据包都有校
9.  *                  验和，由 1 字节模块 ID、1 字节数据头，1 字节二级 ID，6 字节数据和 1 字节校验
10. *                  和构成，因此，数据包长度为 10 字节，数据包的数据容量为 6
11. *              (2) 模块 ID 的最高位 bit7 为 0，数据头、二级 ID、数据以及校验和的最高位为 1，所有包
12. *                  的数据头依次包含数据的最高位，如数据头的 bit0 为二级 ID 的 bit7，数据头的 bit1
13. *                  为数据 1 的 bit7，数据头的 bit2 为数据 2 的 bit7，数据头的 bit6 为数据 6 的 bit7
14. * 注    意:
15. *********************************************************************************************
16. * 取代版本:
17. * 作    者:
18. * 完成日期:
19. * 修改内容:
20. * 修改文件:
21. *********************************************************************************************/
22.
23. /*********************************************************************************************
24. *                                          包含头文件
25. *********************************************************************************************/
26. #include "PackUnpack.h"
27.
28. /*********************************************************************************************
29. *                                            宏定义
30. *********************************************************************************************/
31.
32. /*********************************************************************************************
33. *                                         枚举结构体定义
34. *********************************************************************************************/
35.
36. /*********************************************************************************************
37. *                                           内部变量
38. *********************************************************************************************/
39. //以下 4 个参数在打包和解包时使用
40. static StructPackType s_ptPack;        //数据包，1 字节模块 ID，1 字节数据头，1 字节二级 ID，
                                                                     6 字节数据，1 字节校验和
41. static u8       s_iPackLen;            //数据包长度，用来判断数据长度是否为 10，不为 10 则是
                                                                                       错误包
42. static u8       s_iGotPackId;          //获取到 ID 的标志
43. static u8       s_iRestByteNum;        //剩余字节数
```

```
44.
45. /*********************************************************************************
46. *                                   内部函数声明
47. *********************************************************************************/
48. static  void  PackWithCheckSum(u8* pPack);          //带校验和的数据打包
49. static  u8    UnpackWithCheckSum(u8* pPack);        //带校验和的数据解包
50.
51. /*********************************************************************************
52. *                                   内部函数实现
53. *********************************************************************************/
54.
55. /*********************************************************************************
56. * 函数名称: PackWithCheckSum
57. * 函数功能: 带校验和的数据打包
58. * 输入参数: 待打包的数据 pPack 的首地址
59. * 输出参数: 打包好的数据 pPack 的首地址
60. * 返 回 值: void
61. * 创建日期: 2021 年 03 月 01 日
62. * 注    意: 如数据头的 bit0 为二级 ID 的 bit7，数据头的 bit1 为数据 1 的 bit7，数据头的 bit2 为
63. *           数据 2 的 bit7，数据头的 bit6 为数据 6 的 bit7
64. *********************************************************************************/
65. static void PackWithCheckSum(u8* pPack)
66. {
67.   u8  i;
68.   u8  dataHead;                //数据头，在数据包的第 2 个位置，即 ModuleID 之后
69.   u8  checkSum;                //校验和，在数据包的最后一个位置
70.
71.   checkSum = *(pPack);         //取出 ModuleID，加到校验和
72.   dataHead = 0;                //数据头清零
73.
74.   for(i = 8; i > 1; i --)
75.   {
76.     //数据头左移，后面数据的最高位位于 dataHead 的左边，如数据 6 的最高位位于 dataHead 的 bit6
77.     dataHead <<= 1;
78.
79.     //取出原始数据的最高位，与 dataHead 相或
80.     dataHead |= (*(pPack + i) & 0x80) >> 7;
81.
82.     //对数据进行最高位置 1 操作
83.     *(pPack + i) = *(pPack + i) | 0x80;
84.
85.     //数据加到校验和
86.     checkSum += *(pPack+i);
87.   }
88.
89.   //数据头在数据包的第二个位置，仅次于包头，数据头的最高位也要置为 1
90.   *(pPack+1) = dataHead | 0x80;
91.
92.   //将数据头加到校验和
93.   checkSum += *(pPack+1);
94.
95.   //校验和的最高位也要置为 1
```

```
96.   *(pPack+9) = checkSum | 0x80;
97. }
98.
99. /*********************************************************************************
100. * 函数名称: UnpackWithCheckSum
101. * 函数功能: 带校验和的数据解包
102. * 输入参数: 待解包的数据 pPack 的首地址，打包后的数据长度
103. * 输出参数: 解包好的数据 pPack 的首地址
104. * 返 回 值: 0-解包不成功，1-解包成功
105. * 创建日期: 2021 年 03 月 01 日
106. * 注    意:
107. *********************************************************************************/
108. static u8  UnpackWithCheckSum(u8* pPack)
109. {
110.   u8  i;
111.   u8  dataHead;                      //数据头，在数据包的第 2 个位置，即 ModuleID 之后
112.   u8  checkSum;                      //校验和，在数据包的最后一个位置
113.
114.   checkSum  = *(pPack);              //取出模块 ID，加到校验和
115.
116.   dataHead = *(pPack + 1);           //取出数据包的数据头，赋给 dataHead
117.   checkSum += dataHead;              //将数据头加到校验和
118.
119.   for(i = 2; i < 9; i++)
120.   {
121.     checkSum += *(pPack + i);        //将数据依次加到校验和
122.
123.     //还原二级 ID 和 6 位数据
124.     *(pPack + i) = (*(pPack + i) & 0x7f) | ((dataHead & 0x1) << 7);
125.
126.     dataHead >>= 1;                  //数据头右移一位
127.   }
128.
129.   //判断模块 ID、数据头、二级 ID 和数据求和的结果（低七位）是否与校验和的低七位相等，如果
                                                                            不相等，则返回 0
130.   if((checkSum & 0x7f ) != ((*(pPack + 9)) & 0x7f))
131.   {
132.     return 0;
133.   }
134.
135.   return 1;
136. }
137.
138. /*********************************************************************************
139. *                                API 函数实现
140. *********************************************************************************/
141.
142. /*********************************************************************************
143. * 函数名称: InitPackUnpack
144. * 函数功能: 初始化该模块，其余参数均默认为 0
145. * 输入参数: void
146. * 输出参数: void
```

```
147. * 返 回 值: void
148. * 创建日期: 2021 年 03 月 01 日
149. * 注    意:
150. *********************************************************************************/
151. void  InitPackUnpack(void)
152. {
153.   i16 i;
154.
155.   s_ptPack.packModuleId  = 0;         //s_ptPack 的模块 ID 默认为 0, 即未使用到的 ID
156.   s_ptPack.packHead      = 0;         //s_ptPack 的数据头默认为 0
157.   s_ptPack.packSecondId  = 0;         //s_ptPack 的二级 ID 默认为 0
158.
159.   for(i = 0; i < 6; i++)
160.   {
161.     s_ptPack.arrData[i]   = 0;        //s_ptPack 的 6 个数据默认为 0
162.   }
163.
164.   s_ptPack.checkSum  = 0;             //s_ptPack 的校验和默认为 0
165.
166.   s_iPackLen              = 0;        //数据包的长度默认为 0
167.   s_iGotPackId            = 0;        //获取到数据包 ID 标志默认为 0, 即尚未获取到有效模块 ID
168.   s_iRestByteNum          = 0;        //剩余的字节数默认为 0
169. }
170.
171. /*********************************************************************************
172. * 函数名称: PackData
173. * 函数功能: 对数据进行打包
174. * 输入参数: pPT, 待打包的数据首地址
175. * 输出参数: pPT, 打包好的数据首地址
176. * 返 回 值: valid, 1-打包成功, 0-打包失败
177. * 创建日期: 2021 年 03 月 01 日
178. * 注    意:
179. *********************************************************************************/
180. u8  PackData(StructPackType* pPT)
181. {
182.   u8 valid = 0;
183.
184.   if(pPT->packModuleId < 0x80)               //包 ID 必须在 0x00-0x7F 之间
185.   {
186.     valid = 1;    //表示模块 ID 是合法的
187.     PackWithCheckSum((u8 *)pPT);
188.   }
189.
190.   return(valid);
191. }
192.
193. /*********************************************************************************
194. * 函数名称: UnPackData
195. * 函数功能: 对数据进行解包, 返回 1 表示解析到一个有效包, 此时通过调用 GetUnPackRslt 函数将
196. *           数据包取走, 否则新数据会覆盖解析好的数据包
197. * 输入参数: data, 待解包的数据
198. * 输出参数: void
```

```
199. * 返 回 值: 是否解包成功，1-解包成功，0-解包失败
200. * 创建日期: 2021 年 03 月 01 日
201. * 注    意:
202. *********************************************************************************/
203. u8  UnPackData(u8 data)
204. {
205.   u8 findPack = 0;
206.   u8* pBuf;
207.
208.   pBuf = (u8*)&s_ptPack;           //pBuf 指向 s_ptPack，即 pBuf 和 s_ptPack 的值相等
209.
210.   if(s_iGotPackId)                 //已经接收到包 ID
211.   {
212.     if(0x80 <= data)               //非模块 ID 必须大于或等于 0X80
213.     {
214.       //数据包中的非模块 ID 从第二字节开始存储，因为第一字节是模块 ID
215.       pBuf[s_iPackLen] = data;   //赋给 pBuf[s_iPackLen]，也相当于赋给 s_ptPack 中对应的成员
216.       s_iPackLen++;                //包长自增
217.       s_iRestByteNum--;            //剩余字节数自减
218.
219.       if (0 >= s_iRestByteNum && 10 == s_iPackLen)
220.       {
221.         findPack = UnpackWithCheckSum(pBuf);//接收到完整数据包后尝试解包
222.         s_iGotPackId = 0;                    //清除获取到包 ID 标志，即重新判断下一个数据包
223.       }
224.     }
225.     else
226.     {
227.       s_iGotPackId = 0;              //表示出错
228.     }
229.   }
230.   else if( data < 0x80 )             //当前的数据为包 ID
231.   {
232.     s_iRestByteNum  = 9;             //剩余的包长，即打包好的包长减去 1
233.     s_iPackLen      = 1;             //尚未接收到包 ID，即表示包长为 1
234.     s_ptPack.packModuleId = data; //数据包的 ID
235.     s_iGotPackId    = 1;             //表示已经接收到包 ID
236.   }
237.
238.   return findPack;          //如果获取到完整的数据包，并解包成功，则 findPack 为 1；否则为 0
239. }
240.
241. /*********************************************************************************
242. * 函数名称: GetUnPackRslt
243. * 函数功能: 获取解包结果，主要看 packModuleId、packSecondId 和 arrData[0]-[5]
244. * 输入参数: void
245. * 输出参数: void
246. * 返 回 值: 解包后的结果，packModuleId+packHead+packSecondId+arrData[0]-[5]+checkSum，主
             要看 arrData[0]-[5]
247. * 创建日期: 2021 年 03 月 01 日
248. * 注    意:
249. *********************************************************************************/
```

```
250. StructPackType  GetUnPackRslt(void)
251. {
252.   return(s_ptPack);
253. }
```

步骤 4：完善 App.c 文件

将程序清单 13-3 中的代码输入 App.c 文件中。下面按照顺序对部分语句进行解释。

（1）第 47 至 68 行代码：PrintfDataAfterPack 函数用于打印打包后的数据包。

（2）第 70 至 91 行代码：PrintfDataAfterUnpack 函数用于打印解包后的数据。

（3）第 114 至 116 代码：数组 arrDataBeforeUnpack 保存了一系列数据，包括两个 10 字节 PCT 通信协议数据包，这两个数据包之间还有 2 字节无效数据，分别为 0x22 和 0x41。

（4）第 123 至 136 行代码：分两步对第 1 个数据包进行打包处理，第一步先准备好待打包数据；第二步通过 PackData 函数对待打包数据进行打包处理，如果打包成功，即 PackData 函数返回值为 1，则通过调用 PrintfDataAfterPack 函数打印打包后的结果。

（5）第 138 至 151 行代码：对第 2 个数据包进行打包处理，与第 1 个数据包的处理步骤类似。

（6）第 153 至 162 行代码：数组 arrDataBeforeUnpack 有 100 个元素，使用 for 循环语句，通过逐个调用 UnPackData 函数对这些元素进行解包处理，如果该函数的返回值大于 0，即获取到完整的包，则通过调用 GetUnPackRslt 函数获取解包结果，同时，调用 PrintfDataAfterUnpack 函数打印解包后的结果。

程序清单 13-3

```
1.   /*********************************************************************************
2.   * 模块名称: App.c
3.   * 摘    要: 测试 PCT 协议
4.   * 当前版本: 1.0.0
5.   * 作    者: Leyutek(COPYRIGHT 2018 - 2021 Leyutek. All rights reserved.)
6.   * 完成日期: 2021 年 03 月 01 日
7.   * 内    容:
8.   * 注    意:
9.   *********************************************************************************
10.  * 取代版本:
11.  * 作    者:
12.  * 完成日期:
13.  * 修改内容:
14.  * 修改文件:
15.  *********************************************************************************/
16.
17.  /*********************************************************************************
18.  *                                包含头文件
19.  *********************************************************************************/
20.  #include <stdio.h>
21.  #include <stdlib.h>
22.  #include "PackUnpack.h"
23.  #include "DataType.h"
24.
```

```
25. /*********************************************************************************
26. *                                    宏定义
27. *********************************************************************************/
28.
29. /*********************************************************************************
30. *                                 枚举结构体定义
31. *********************************************************************************/
32.
33. /*********************************************************************************
34. *                                   内部变量
35. *********************************************************************************/
36.
37. /*********************************************************************************
38. *                                 内部函数声明
39. *********************************************************************************/
40. static void PrintfDataAfterPack(u8* pData);                       //打印打包后的数据包
41. static void PrintfDataAfterUnpack(u8* pData, i16 packLen);  //打印解包后的数据
42.
43. /*********************************************************************************
44. *                                 内部函数实现
45. *********************************************************************************/
46.
47. /*********************************************************************************
48. * 函数名称: PrintfDataAfterPack
49. * 函数功能: 打印打包后的数据包，第一个为 ID，第二个为包头，第三个开始为数据，最后一个为校验和
50. * 输入参数: pData-待打印数据包的首地址，packLen-打印数据的个数
51. * 输出参数: void
52. * 返 回 值: void
53. * 创建日期: 2021 年 03 月 01 日
54. * 注    意:
55. *********************************************************************************/
56. static void PrintfDataAfterPack(u8* pData)
57. {
58.   static  u8  s_iCnt = 1;
59.   i16     i;
60.
61.   printf("Packet NO.%d after packing:\n", s_iCnt);
62.   s_iCnt++;
63.
64.   for( i = 0; i < 10; i++ )
65.   {
66.     printf("pData[%d] = %x \n", i, pData[i]);
67.   }
68. }
69.
70. /*********************************************************************************
71. * 函数名称: PrintfDataAfterUnpack
72. * 函数功能: 打印解包后的数据，第一个为 ID，第二个开始为数据
```

```
73. * 输入参数：pData-待打印数据包的首地址，packLen-打印数据的个数
74. * 输出参数：void
75. * 返 回 值：void
76. * 创建日期：2021 年 03 月 01 日
77. * 注    意：
78. ************************************************************************************/
79. static void PrintfDataAfterUnpack(u8* pData, i16 packLen)
80. {
81.   static  u8  s_iCnt = 1;
82.   i16 i;
83.
84.   printf("Data NO.%d after unpacking:\n", s_iCnt);
85.   s_iCnt++;
86.
87.   for( i = 0; i < packLen; i++ )
88.   {
89.     printf("pData[%d] = %x \n", i, pData[i]);
90.   }
91. }
92.
93. /************************************************************************************
94. *                                     API 函数实现
95. ************************************************************************************/
96.
97. /************************************************************************************
98. * 函数名称：main
99. * 函数功能：主函数，用于测试 PackUnpack 模块
100. * 输入参数：void
101. * 输出参数：void
102. * 返 回 值：void
103. * 创建日期：2021 年 03 月 01 日
104. * 注    意：
105. ************************************************************************************/
106. void main( void )
107. {
108.    u8  i = 0;
109.    u8  packGot;                          //获取到数据包标志，1-获取到数据包，0-未获取到
110.
111.    StructPackType  ptDataBeforePack1;    //打包前的数据包 1
112.    StructPackType  ptDataBeforePack2;    //打包前的数据包 2
113.
114.    u8  arrDataBeforeUnpack[100] = {0x70, 0xa0, 0x82, 0x92, 0xb4, 0xd6, 0xf8, 0x90, 0x80, 0xb6,
115.                                    0x22, 0x41, 0x71, 0xa0,
116.                                    0x82, 0x91, 0xb3, 0xd5, 0xf7, 0x99, 0x80, 0xbc};
117.
118.    StructPackType  ptDataAfterUnPack;    //解包后的数据包，即第一个为 ID，第二个开始为数据
119.
120.    InitPackUnpack();                     //初始化
```

```
121.
122.
123.   //对第 1 个数据包进行打包
124.   ptDataBeforePack1.packModuleId = MODULE_WAVE1;
125.   ptDataBeforePack1.packSecondId = DAT_WAVE1_WDATA;
126.   ptDataBeforePack1.arrData[0]   = 0x12;
127.   ptDataBeforePack1.arrData[1]   = 0x34;
128.   ptDataBeforePack1.arrData[2]   = 0x56;
129.   ptDataBeforePack1.arrData[3]   = 0x78;
130.   ptDataBeforePack1.arrData[4]   = 0x90;
131.   ptDataBeforePack1.arrData[5]   = 0x00;
132.
133.   if(1 == PackData(&ptDataBeforePack1))              //调用打包函数来打包数据
134.   {
135.     PrintfDataAfterPack((u8*)&ptDataBeforePack1);   //打印打包
136.   }
137.
138.   //对第 2 个数据包进行打包
139.   ptDataBeforePack2.packModuleId = MODULE_WAVE2;
140.   ptDataBeforePack2.packSecondId = DAT_WAVE2_WDATA;
141.   ptDataBeforePack2.arrData[0]   = 0x11;
142.   ptDataBeforePack2.arrData[1]   = 0x33;
143.   ptDataBeforePack2.arrData[2]   = 0x55;
144.   ptDataBeforePack2.arrData[3]   = 0x77;
145.   ptDataBeforePack2.arrData[4]   = 0x99;
146.   ptDataBeforePack2.arrData[5]   = 0x00;
147.
148.   if(1 == PackData(&ptDataBeforePack2))                        //调用打包函数来打包数据
149.   {
150.     PrintfDataAfterPack((u8*)&ptDataBeforePack2);              //打印打包
151.   }
152.
153.   //依次对数据进行解包
154.   for(i = 0; i < 100; i++)
155.   {
156.     packGot = UnPackData(arrDataBeforeUnpack[i]);
157.     if(packGot > 0)                                            //获取到完整的包
158.     {
159.       ptDataAfterUnPack = GetUnPackRslt();                     //获取到解包后的结果
160.       PrintfDataAfterUnpack((u8*)&ptDataAfterUnPack, 10);      //打印解包后的结果
161.     }
162.   }
163.
164.   system("pause");
165. }
```

步骤 5：项目编译和运行

最后，按 F5 键编译并运行程序，在弹出的控制台窗口中，可以看到如图 13-12 所示的运

行结果，说明实验成功。为了与实验结果进行对比，可以手动计算打包和解包结果。

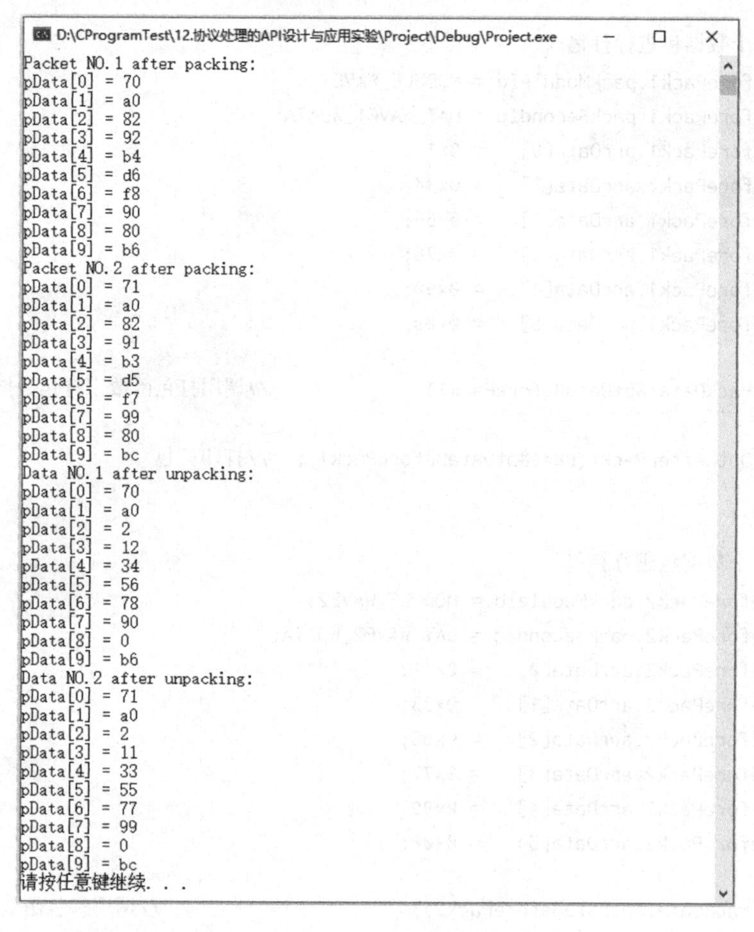

图 13-12　本章项目运行结果

本 章 任 务

使用智能小车作为从机模块，模块 ID 为 0x13（MODULE_CAR）。智能小车需要传输采集到的摄像头数据到主机（如手机），这里规定摄像头数据包的二级 ID 为 0x02（DAT_CAR_CAMERA），假设摄像头数据包中的数据 1～数据 6 依次为 0x12、0x23、0x34、0x78、0x89、0x9A，基于本章实验的 PackUnpack 模块，对待打包的摄像头数据进行打包处理，并通过 printf 函数打印打包结果，再将打包结果进行解包处理，通过 printf 函数打印解包结果。

本 章 习 题

1．什么是 PCT 通信协议？简述 PCT 通信协议的数据包格式。

2．PackUnpack 模块都有哪些 API 函数？简述每个 API 函数的功能。

第 14 章　模拟从机命令接收与数据发送

本章实验通过模拟的方式实现从机命令接收与数据发送，命令接收是指从机对接收到的命令进行接收、解包和处理，数据发送是指从机对数据进行打包和发送。最后，在 App.c 文件中对从机命令接收与数据发送进行测试验证。

14.1　实验内容

使用第 13 章中的 PTC 通信协议，模拟从机命令接收与数据发送，如图 14-1 所示。这里可以将主机假设为计算机，计算机用于发送命令和接收数据，并将接收到的数据通过屏幕显示出来。将从机假设为单片机，单片机用于接收计算机发送的命令，当接收到生成正弦波（wave1）命令时，单片机向计算机发送正弦波数据；当接收到生成方波（wave2）命令时，向计算机发送方波数据。

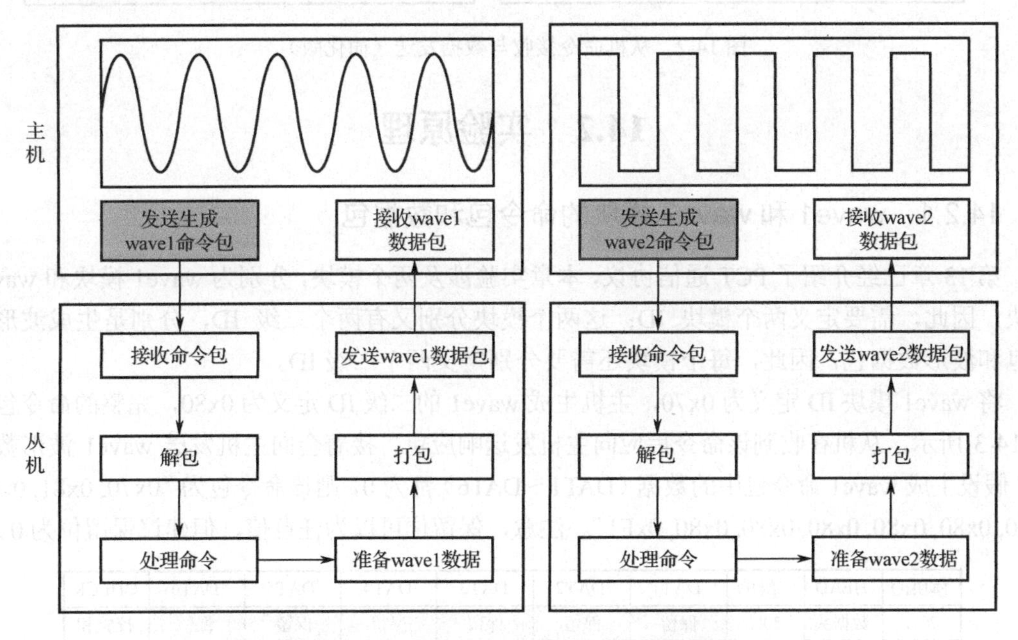

图 14-1　从机命令接收与数据发送（原始版）

模拟从机命令接收与数据发送的具体流程如下：①当从机收到主机发送的生成 wave1 命令包时，从机先进行解包处理，然后处理命令，准备 wave1 数据，并对 wave1 数据进行打包，最后，将打包好的 wave1 数据包发送给主机；主机将接收到的 wave1 数据包进行解包后，就可以在屏幕上显示正弦波。②当从机收到主机发送的生成 wave2 命令包时，从机先进行解包处理，然后处理命令，准备 wave2 数据，并对 wave2 数据进行打包，最后，将打包好的 wave2 数据包发送给主机；主机将接收到的 wave2 数据包进行解包后，就可以在屏幕上显示方波。

为了简化实验，wave1 模块波形数据固定为 0x11，wave2 模块波形数据固定为 0x22，如图 14-2 所示，wave1 和 wave2 均为一条直线。

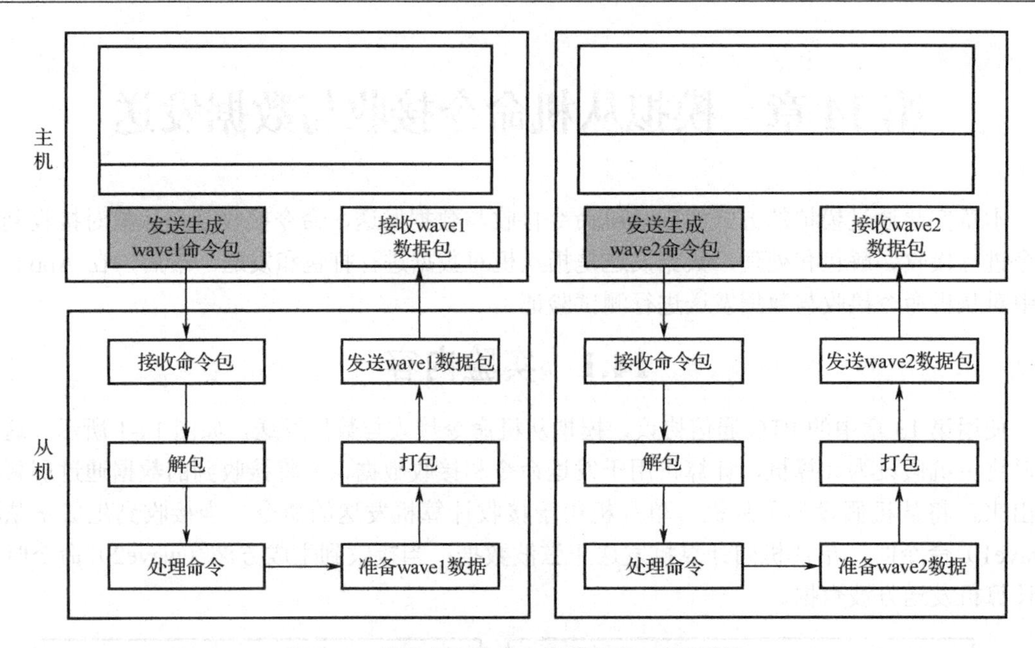

图 14-2　从机命令接收与数据发送（简化版）

14.2　实验原理

14.2.1　wave1 和 wave2 模块的命令包和数据包

第 13 章已经介绍了 PCT 通信协议，本章实验涉及两个模块，分别为 wave1 模块和 wave2 模块，因此，需要定义两个模块 ID；这两个模块分别又有两个二级 ID，分别是生成波形命令包和波形数据包，因此，每个模块还需要分别定义两个二级 ID。

将 wave1 模块 ID 定义为 0x70，主机生成 wave1 的二级 ID 定义为 0x80，完整的命令包如图 14-3 所示，从机在收到该命令后应向主机发送响应包，接着会向主机发送 wave1 波形数据包。假设生成 wave1 命令包中的数据（DAT1～DAT6）都为 0，则该命令包为“0x70, 0x81, 0x80, 0x80, 0x80, 0x80, 0x80, 0x80, 0x80, 0xF1”。注意，保留位可以为任意值，但建议保留位为 0。

模块ID	HEAD	二级ID	DAT1	DAT2	DAT3	DAT4	DAT5	DAT6	CHECK
70H	数据头	80H	保留	保留	保留	保留	保留	保留	校验和

图 14-3　wave1 模块生成 wave1 命令包

将 wave2 模块 ID 定义为 0x71，主机生成 wave2 的二级 ID 定义为 0x80，完整的命令包如图 14-4 所示，从机在收到该命令后应向主机发送响应包，接着会向主机发送 wave2 波形数据包。假设生成 wave2 命令包中的数据（DAT1～DAT6）都为 0，则该命令包为“0x71, 0x81, 0x80, 0x80, 0x80, 0x80, 0x80, 0x80, 0x80, 0xF2”。注意，不同模块的二级 ID 可以相同，因此，wave1 模块和 wave2 模块的二级 ID 都为 0x80，符合 PCT 通信协议。

模块ID	HEAD	二级ID	DAT1	DAT2	DAT3	DAT4	DAT5	DAT6	CHECK
71H	数据头	80H	保留	保留	保留	保留	保留	保留	校验和

图 14-4　wave2 模块生成 wave2 命令包

将从机向主机发送 wave1 的数据包的二级 ID 定义为 0x02，完整的数据包如图 14-5 所示。如果 DAT1～DAT6 均为 0x11，则该数据包为“0x70,0x80,0x82,0x91,0x91,0x91,0x91,0x91,0x91,0xD8”。

模块ID	HEAD	二级ID	DAT1	DAT2	DAT3	DAT4	DAT5	DAT6	CHECK
70H	数据头	02H	波形数据1	波形数据2	波形数据3	波形数据4	波形数据5	保留	校验和

图 14-5　wave1 模块波形数据包

注意，DAT1～DAT5 为波形数据 1～5，因为只有 5 字节波形数据，所以 DAT6 保留。

将从机向主机发送 wave2 的数据包的二级 ID 定义为 0x02，完整的数据包如图 14-6 所示。如果 DAT1～DAT6 均为 0x22，则该数据包为“0x71,0x80,0x82,0xA2,0xA2,0xA2,0xA2,0xA2,0xA2,0xBF”。

模块ID	HEAD	二级ID	DAT1	DAT2	DAT3	DAT4	DAT5	DAT6	CHECK
71H	数据头	02H	波形数据1	波形数据2	波形数据3	波形数据4	波形数据5	保留	校验和

图 14-6　wave2 模块波形数据包

注意，这些模块 ID 和二级 ID 用户可以自定义，但必须遵循 PCT 通信协议的规则。

14.2.2　新增 wave1 和 wave2 模块通信协议

PCT 通信协议不仅可以应用在人体生理参数监测系统上，还可以应用于其他项目或产品，下面以 wave1 和 wave2 模块为例，介绍如何在 C 文件中使用 PCT 通信协议。（PCT 通信协议还可以通过其他语言实现，如 C++、C#、Java、MATLAB、Python、VHDL 和 Verilog HDL 等，可参阅“卓越工程师培养系列”其他教材。）

在 PackUnpack.h 文件的枚举定义中添加新增的模块 ID，新增的模块 ID 不能与已有的模块 ID 重复，且新增的模块 ID 必须在 0x00～0x7F 之间，然后添加新增二级 ID 的枚举定义。注意，二级 ID 的 0x00～0x7F 约定为从机到主机的数据包二级 ID，0x80～0xFF 约定为主机到从机的命令包二级 ID。

将这两个模块的模块 ID 和二级 ID 增加到代码中，具体方法如下。

（1）由于 wave1 模块和 wave2 模块的模块 ID 分别为 0x70（MODULE_WAVE1）和 0x71（MODULE_WAVE2），因此，首先在模块 ID 的枚举定义中新增这两个模块 ID，如程序清单 14-1 所示。注意，新增一个模块，只需要将其模块 ID 作为枚举元素添加到枚举 EnumPackID 中即可，例如，wave1 模块的模块 ID（MODULE_WAVE1）为枚举 EnumPackID 中的枚举元素。

程序清单 14-1

```
1.   //枚举定义，定义模块 ID，0x00-0x7F，不可以重复
2.   typedef enum
3.   {
4.     MODULE_SYS      = 0x01,          //系统信息
5.
6.     MODULE_WAVE1    = 0x70,          //波形 1 信息
7.     MODULE_WAVE2    = 0x71,          //波形 2 信息
8.
9.     MAX_MODULE_ID   = 0x80
```

```
10. }EnumPackID;
```

（2）添加完模块 ID，进一步添加二级 ID。wave1 模块包含生成 wave1 命令包和波形数据包，分别为 0x80 和 0x02，wave2 模块也包含生成 wave2 命令包和波形数据包，分别为 0x80 和 0x02，因此，在 PackUnpack.h 文件的“枚举结构体定义”区的最后，添加两个模块的二级 ID，如程序清单 14-2 所示。注意，新增一个模块，就需要声明一个枚举，该枚举包含的枚举元素即为该模块的二级 ID，例如，wave1 模块的二级 ID（DAT_WAVE1_WDATA 和 CMD_GEN_WAVE1）为枚举 EnumWave1SecondID 中的枚举元素。

程序清单 14-2

```
1.  //wave1 模块的二级 ID
2.  typedef enum
3.  {
4.    DAT_WAVE1_WDATA = 0x02,          //波形 1 的波形数据
5.
6.    CMD_GEN_WAVE1   = 0x80,          //生成波形 1 命令
7.  }EnumWave1SecondID;
8.
9.  //wave2 模块的二级 ID
10. typedef enum
11. {
12.   DAT_WAVE2_WDATA = 0x02,          //波形 2 的波形数据
13.
14.   CMD_GEN_WAVE2   = 0x80,          //生成波形 2 命令
15. }EnumWave2SecondID;
```

从上述流程可以看出，在 PackUnpack 模块中新增通信协议，实质上就是在 PackUnpack.h 文件中新增模块 ID（枚举元素的形式）和二级 ID（声明一个新的枚举），而无须更改 PackUnpack.c 文件。

14.2.3 从机命令接收流程说明

图 14-7 是从机命令接收流程图，具体的接收流程如下：①启动 5s 定时器；②判断 5s 计数是否溢出，如果未产生溢出，则继续判断；③如果产生溢出，则处理主机命令，并对主机命令进行解包；④判断解包结果，如果接收到完整包，则从解包结果中获取解包后的命令；⑤判断是否收到生成 wave1 命令，如果收到该命令，则启动发送 wave1 定时器，并关闭发送 wave2 定时器，同时还要发送响应包；⑥如果未收到生成 wave1 命令，则继续判断是否收到生成 wave2 命令，如果收到该命令，则启动发送 wave2 定时器，并关闭发送 wave1 定时器，同时发送响应包；⑦如果既未收到生成 wave1 命令，也未收到生成 wave2 命令，则关闭发送波形定时器，继续判断 5s 计数是否溢出。

14.2.4 从机数据发送流程说明

图 14-8 是从机数据发送流程图，由于从机发送 wave1 和 wave2 波形数据的流程相同，这里仅介绍从机发送 wave1 的流程，具体如下：①启动生成 wave1 的 200ms 定时器；②判断 200ms 计数是否溢出，如果未产生溢出，则继续判断；③如果产生溢出，则发送 wave1 到主机。

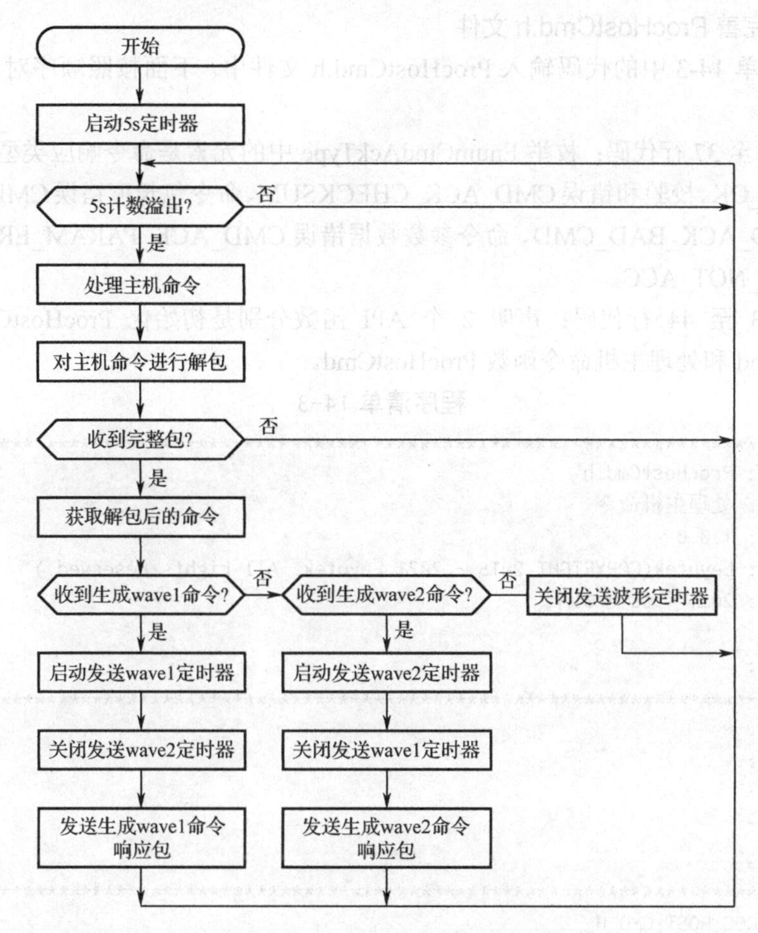

图 14-7　从机命令接收流程图

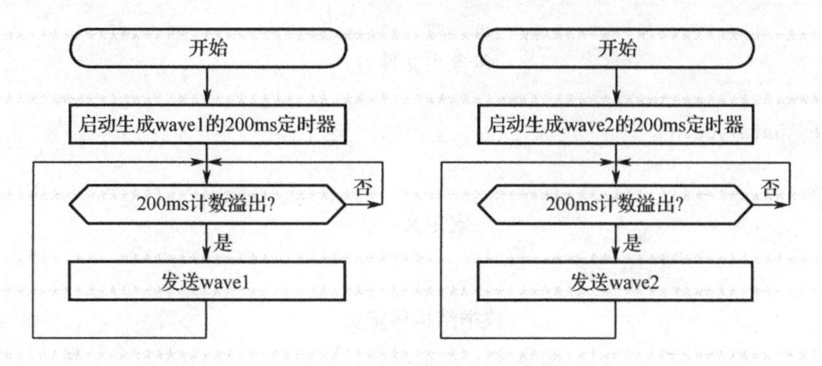

图 14-8　从机数据发送流程图

14.3　实验步骤

步骤 1：复制原始项目

首先，将本书配套资料包的“03.例程资料\Material\13.模拟从机命令接收与数据发送实验”文件夹复制到 CProgramTest 文件夹中，然后，双击运行“D:\CProgramTest\13.模拟从机命令接收与数据发送实验\Project”文件夹中的 Project.sln 文件。

步骤 2：完善 ProcHostCmd.h 文件

将程序清单 14-3 中的代码输入 ProcHostCmd.h 文件中。下面按照顺序对部分语句进行解释。

（1）第 30 至 37 行代码：枚举 EnumCmdAckType 中的元素是命令响应类型，包括命令成功 CMD_ACK_OK、校验和错误 CMD_ACK_CHECKSUM、命令包长度错误 CMD_ACK_LEN、无效命令 CMD_ACK_BAD_CMD、命令参数数据错误 CMD_ACK_PARAM_ERR 和命令不接受 CMD_ACK_NOT_ACC。

（2）第 43 至 44 行代码：声明 2 个 API 函数分别是初始化 ProcHostCmd 模块函数 InitProcHostCmd 和处理主机命令函数 ProcHostCmd。

程序清单 14-3

```
1.  /*********************************************************************************
2.  * 模块名称: ProcHostCmd.h
3.  * 摘    要: 处理主机命令
4.  * 当前版本: 1.0.0
5.  * 作    者: Leyutek(COPYRIGHT 2018 - 2021 Leyutek. All rights reserved.)
6.  * 完成日期: 2021 年 03 月 01 日
7.  * 内    容:
8.  * 注    意:
9.  *********************************************************************************
10. * 取代版本:
11. * 作    者:
12. * 完成日期:
13. * 修改内容:
14. * 修改文件:
15. *********************************************************************************/
16. #ifndef _PROC_HOST_CMD_H_
17. #define _PROC_HOST_CMD_H_
18.
19. /*********************************************************************************
20. *                                  包含头文件
21. *********************************************************************************/
22. #include "DataType.h"
23.
24. /*********************************************************************************
25. *                                   宏定义
26. *********************************************************************************/
27. /*********************************************************************************
28. *                                 枚举结构体定义
29. *********************************************************************************/
30. typedef enum{
31.   CMD_ACK_OK,           //0 命令成功
32.   CMD_ACK_CHECKSUM,     //1 校验和错误
33.   CMD_ACK_LEN,          //2 命令包长度错误
34.   CMD_ACK_BAD_CMD,      //3 无效命令
35.   CMD_ACK_PARAM_ERR,    //4 命令参数数据错误
36.   CMD_ACK_NOT_ACC       //5 命令不接受
37. }EnumCmdAckType;
38.
39.
```

```
40.  /*********************************************************************************************
41.  *                                     API 函数声明
42.  *********************************************************************************************/
43.  void  InitProcHostCmd(void);     //初始化 ProcHostCmd 模块
44.  void  ProcHostCmd(u8 recData);  //处理主机命令
45.
46.  #endif
```

步骤 3：完善 ProcHostCmd.c 文件

将程序清单 14-4 中的代码输入 ProcHostCmd.c 文件中。下面按照顺序对部分语句进行解释。

（1）第 59 至 91 行代码：TimerISRForWave1 和 TimerISRForWave2 函数均为定时器的中断服务函数。其中，TimerISRForWave1 函数用于对 wave1 波形数据进行打包，并发送 wave1 数据包到主机，其中 wave1 的波形数据保存在数组 arrWave1Data 中，该数组有 5 个元素，每个元素均为 0x11。SendWave1ToHost 函数用于发送数据到主机。TimerISRForWave2 函数的功能类似，不再赘述。

（2）第 93 至 123 行代码：OnGenWave1 函数是生成 wave1 命令的响应函数。timeSetEvent 函数用于启动定时器，并返回该定时器的标识。timeKillEvent 函数用于关闭生成 wave2 的定时器。生成 wave2 命令的响应函数 OnGenWave2 与生成 wave1 的响应函数类似。

（3）第 142 至 188 行代码：ProcHostCmd 函数用于处理主机发送到从机的命令，while 循环语句的表达式是 UnPackData 函数的返回值，当该函数的返回值为真，即接收到完整包时，通过 GetUnPackRslt 函数获取解包后的结果，并将其赋值给 pack，pack 是一个结构体变量，根据 pack 的 packModuleId 成员可以判断接收到的命令是生成 wave1 还是生成 wave2 命令。如果接收到的是生成 wave1 命令，则执行生成 wave1 的响应函数 OnGenWave1，同时通过调用 SendAckPack 函数发送响应包；如果接收到的是生成 wave2 命令，则执行生成 wave2 的响应函数 OnGenWave2，同时通过调用 SendAckPack 函数发送响应包；如果接收到的既不是生成 wave1 命令，也不是生成 wave2 命令，则通过 timeKillEvent 函数关闭生成 wave1 和 wave2 的定时器。

程序清单 14-4

```
1.   /*********************************************************************************************
2.   * 模块名称：ProcHostCmd.c
3.   * 摘    要：处理主机命令
4.   * 当前版本：1.0.0
5.   * 作    者：Leyutek(COPYRIGHT 2018 - 2021 Leyutek. All rights reserved.)
6.   * 完成日期：2021 年 03 月 01 日
7.   * 内    容：
8.   * 注    意：
9.   *********************************************************************************************
10.  * 取代版本：
11.  * 作    者：
12.  * 完成日期：
13.  * 修改内容：
14.  * 修改文件：
15.  *********************************************************************************************/
16.
17.  /*********************************************************************************************
```

```
18.  *                                    包含头文件
19.  *********************************************************************************/
20.  #include <stdio.h>
21.  #include <windows.h>
22.
23.  #include "ProcHostCmd.h"
24.  #include "PackUnpack.h"
25.  #include "SendDataToHost.h"
26.
27.  /*********************************************************************************
28.  *                                    宏定义
29.  *********************************************************************************/
30.
31.  /*********************************************************************************
32.  *                                  枚举结构体定义
33.  *********************************************************************************/
34.
35.  /*********************************************************************************
36.  *                                    内部变量
37.  *********************************************************************************/
38.  static unsigned int s_iTimerWave1Id;
39.  static unsigned int s_iTimerWave2Id;
40.
41.  /*********************************************************************************
42.  *                                  内部函数声明
43.  *********************************************************************************/
44.  //定时器的中断服务函数声明，用于生成 wave1
45.  static void __stdcall TimerISRForWave1(unsigned int uTimerID, unsigned int uMsg, unsigned
46.                              long dwUser, unsigned long dw1, unsigned long dw2);
47.  //定时器的中断服务函数声明，用于生成 wave2
48.  static void __stdcall TimerISRForWave2(unsigned int uTimerID, unsigned int uMsg, unsigned
49.                              long dwUser, unsigned long dw1, unsigned long dw2);
50.
51.  //生成 wave1 的响应函数声明
52.  static  u8 OnGenWave1(u8* pMsg);
53.  //生成 wave2 的响应函数声明
54.  static  u8 OnGenWave2(u8* pMsg);
55.
56.  /*********************************************************************************
57.  *                                  内部函数实现
58.  *********************************************************************************/
59.  /*********************************************************************************
60.  * 函数名称：TimerISRForWave1
61.  * 函数功能：定时器的中断服务函数，用于生成 wave1
62.  * 输入参数：*
63.  * 输出参数：void
64.  * 返 回 值：void
65.  * 创建日期：2021 年 03 月 01 日
66.  * 注    意：
67.  *********************************************************************************/
68.  static void __stdcall TimerISRForWave1(unsigned int uTimerID, unsigned int uMsg, unsigned
69.                              long dwUser, unsigned long dw1, unsigned long dw2)
```

```
70. {
71.   u8 arrWave1Data[5] = {0x11, 0x11, 0x11, 0x11, 0x11};
72.   SendWave1ToHost(arrWave1Data);
73.   printf("\n");
74. }
75.
76. /*********************************************************************************
77. * 函数名称: TimerISRForWave2
78. * 函数功能: 定时器的中断服务函数，用于生成 wave2
79. * 输入参数: *
80. * 输出参数: void
81. * 返 回 值: void
82. * 创建日期: 2021 年 03 月 01 日
83. * 注    意:
84. *********************************************************************************/
85. static void __stdcall TimerISRForWave2(unsigned int uTimerID, unsigned int uMsg, unsigned
86.                                 long dwUser, unsigned long dw1, unsigned long dw2)
87. {
88.   u8 arrWave2Data[5] = {0x22, 0x22, 0x22, 0x22, 0x22};
89.   SendWave2ToHost(arrWave2Data);
90.   printf("\n");
91. }
92.
93. /*********************************************************************************
94. * 函数名称: OnGenWave1
95. * 函数功能: 生成 wave1
96. * 输入参数: pMsg
97. * 输出参数: void
98. * 返 回 值: 响应结果
99. * 创建日期: 2021 年 03 月 01 日
100. * 注    意:
101. *********************************************************************************/
102. static  u8 OnGenWave1(u8* pMsg)
103. {
104.   s_iTimerWave1Id = timeSetEvent(200, 1, TimerISRForWave1, pMsg[0], TIME_PERIODIC);
105.   timeKillEvent(s_iTimerWave2Id);
106.   return(CMD_ACK_OK);
107. }
108.
109. /*********************************************************************************
110. * 函数名称: OnGenWave2
111. * 函数功能: 生成 wave2
112. * 输入参数: pMsg
113. * 输出参数: void
114. * 返 回 值: 响应结果
115. * 创建日期: 2021 年 03 月 01 日
116. * 注    意:
117. *********************************************************************************/
118. static  u8 OnGenWave2(u8* pMsg)
119. {
120.   s_iTimerWave2Id = timeSetEvent(200, 1, TimerISRForWave2, pMsg[0], TIME_PERIODIC);
121.   timeKillEvent(s_iTimerWave1Id);
```

```
122.   return(CMD_ACK_OK);
123. }
124.
125. /*********************************************************************************
126. *                                    API 函数实现
127. *********************************************************************************/
128.
129. /*********************************************************************************
130. * 函数名称: InitProcHostCmd
131. * 函数功能: 初始化该模块
132. * 输入参数: void
133. * 输出参数: void
134. * 返 回 值: void
135. * 创建日期: 2021 年 03 月 01 日
136. * 注    意:
137. *********************************************************************************/
138. void  InitProcHostCmd(void)
139. {
140. }
141.
142. /*********************************************************************************
143. * 函数名称: ProcHostCmd
144. * 函数功能: 处理主机发送来的命令
145. * 输入参数: recData
146. * 输出参数: void
147. * 返 回 值: void
148. * 创建日期: 2021 年 03 月 01 日
149. * 注    意:
150. *********************************************************************************/
151. void ProcHostCmd(u8 recData)
152. {
153.   u8 ack;
154.   StructPackType pack;
155.
156.   while(UnPackData(recData))
157.   {
158.     pack = GetUnPackRslt();                  //获取解包结果
159.     switch(pack.packModuleId)
160.     {
161.       case MODULE_WAVE1:
162.         if(CMD_GEN_WAVE1 == pack.packSecondId)
163.         {
164.             ack = OnGenWave1(pack.arrData);
165.             printf("GEN_WAVE1 ACK:\n");
166.             SendAckPack(MODULE_WAVE1, CMD_GEN_WAVE1, ack);      //发送响应包
167.             printf("\n");
168.             printf("------WAVE1------\n");
169.             //printf("ID1-%d\n",s_iTimerWave1Id);
170.          }
171.         break;
172.       case MODULE_WAVE2:
173.         if(CMD_GEN_WAVE2 == pack.packSecondId)
```

```
174.          {
175.              ack = OnGenWave2(pack.arrData);
176.               printf("GEN_WAVE2 ACK:\n");
177.               SendAckPack(MODULE_WAVE2, CMD_GEN_WAVE2, ack);      //发送响应包
178.               printf("\n");
179.               printf("------WAVE2------\n");
180.          }
181.           break;
182.        default:
183.          timeKillEvent(s_iTimerWave1Id);
184.          timeKillEvent(s_iTimerWave2Id);
185.          break;
186.      }
187.   }
188. }
```

步骤 4：完善 SendDataToHost.h 文件

将程序清单 14-5 中的代码输入 SendDataToHost.h 文件中。在 API 函数声明区，声明 4 个函数，分别是初始化 SendDataToHost 模块函数 InitSendDataToHost、发送响应包函数 SendAckPack、发送 wave1 到主机函数 SendWave1ToHost 和发送 wave2 到主机函数 SendWave2ToHost。

程序清单 14-5

```
1.  /*********************************************************************************
2.  * 模块名称: SendDataToHost.h
3.  * 摘    要: 发送数据到主机
4.  * 当前版本: 1.0.0
5.  * 作    者: Leyutek(COPYRIGHT 2018 - 2021 Leyutek. All rights reserved.)
6.  * 完成日期: 2021 年 03 月 01 日
7.  * 内    容:
8.  * 注    意:
9.  *********************************************************************************
10. * 取代版本:
11. * 作    者:
12. * 完成日期:
13. * 修改内容:
14. * 修改文件:
15. *********************************************************************************/
16. #ifndef _SEND_DATA_TO_HOST_H_
17. #define _SEND_DATA_TO_HOST_H_
18.
19. /*********************************************************************************
20. *                                包含头文件
21. *********************************************************************************/
22. #include "DataType.h"
23.
24. /*********************************************************************************
25. *                                 宏定义
26. *********************************************************************************/
27.
28. /*********************************************************************************
29. *                               枚举结构体定义
```

```
30. *****************************************************************************/
31.
32. /*****************************************************************************
33. *                                   API 函数声明
34. *****************************************************************************/
35. void  InitSendDataToHost(void);           //初始化 SendDataToHost 模块
36. void  SendAckPack(u8 moduleId, u8 secondId, u8 ackMsg); //发送响应包
37.
38. void  SendWave1ToHost(u8* pWaveData);    //发送 wave1 到主机，一次性发送 5 个点
39. void  SendWave2ToHost(u8* pWaveData);    //发送 wave2 到主机，一次性发送 5 个点
40.
41. #endif
```

步骤 5：完善 SendDataToHost.c 文件

将程序清单 14-6 中的代码输入 SendDataToHost.c 文件中。下面按照顺序对部分语句进行解释。

（1）第 44 至 67 行代码：SendPackToHost 函数的作用是发送包到主机，由于本章实验是模拟从机发送数据，因此这里的“发送”过程实质上是通过调用 printf 函数打印 wave1 数据包，在打印之前，要先通过调用 PackData 函数对原始数据（包括模块 ID、二级 ID 和 6 字节数据）进行打包。注意，在该函数中，(u8 *)pPackSent 的功能是将结构体类型的指针变量 pPackSent 转换为 u8 类型的指针变量，这样就可以方便地获取该结构体中的每个字节。

（2）第 85 至 108 行代码：SendAckPack 函数的作用是发送响应包到主机，该函数的主要功能是将模块 ID、二级 ID 和 6 字节数据保存到结构体变量 pt 中，然后调用 SendPackToHost 函数，该函数的作用是打包数据，并将打包好的数据包发送到主机，其中，pt 的地址作为该函数的参数。

（3）第 110 至 158 行代码：SendWave1ToHost 函数的作用是发送 wave1 波形数据包到主机，该数据包包含 wave1 波形的 5 个点。SendWave2ToHost 函数的功能与 SendWave1ToHost 函数的功能类似。

程序清单 14-6

```
1.  /*****************************************************************************
2.  * 模块名称: SendDataToHost.c
3.  * 摘    要: 发送数据到主机
4.  * 当前版本: 1.0.0
5.  * 作    者: Leyutek(COPYRIGHT 2018 - 2021 Leyutek. All rights reserved.)
6.  * 完成日期: 2021 年 03 月 01 日
7.  * 内    容:
8.  * 注    意:
9.  ******************************************************************************
10. * 取代版本:
11. * 作    者:
12. * 完成日期:
13. * 修改内容:
14. * 修改文件:
15. *****************************************************************************/
16.
17. /*****************************************************************************
18. *                                   包含头文件
19. *****************************************************************************/
```

```
20.  #include <stdio.h>
21.  #include "SendDataToHost.h"
22.  #include "PackUnpack.h"
23.
24.  /*********************************************************************************
25.  *                                宏定义
26.  *********************************************************************************/
27.
28.  /*********************************************************************************
29.  *                                枚举结构体定义
30.  *********************************************************************************/
31.
32.  /*********************************************************************************
33.  *                                内部变量
34.  *********************************************************************************/
35.
36.  /*********************************************************************************
37.  *                                内部函数声明
38.  *********************************************************************************/
39.  static  void  SendPackToHost(StructPackType* pPackSent);    //发送包到主机
40.
41.  /*********************************************************************************
42.  *                                内部函数实现
43.  *********************************************************************************/
44.  /*********************************************************************************
45.  * 函数名称: SendPackToHost
46.  * 函数功能: 发送包到主机
47.  * 输入参数: packSent
48.  * 输出参数: void
49.  * 返 回 值: void
50.  * 创建日期: 2021 年 03 月 01 日
51.  * 注    意:
52.  *********************************************************************************/
53.  static  void  SendPackToHost(StructPackType* pPackSent)
54.  {
55.    u8  valid;
56.    i16 i = 0;
57.
58.    valid = PackData(pPackSent);     //调用打包函数来打包数据
59.
60.    if(1 == valid)
61.    {
62.      for(i = 0; i < 10; i++)
63.      {
64.        printf("%02x ", *((u8 *)pPackSent + i));
65.      }
66.    }
67.  }
68.
69.  /*********************************************************************************
70.  *                                API 函数实现
71.  *********************************************************************************/
```

```
72.  /*************************************************************************************
73.  * 函数名称: InitSendDataToHost
74.  * 函数功能: 初始化该模块
75.  * 输入参数: void
76.  * 输出参数: void
77.  * 返 回 值: void
78.  * 创建日期: 2021 年 03 月 01 日
79.  * 注    意:
80.  *************************************************************************************/
81.  void  InitSendDataToHost(void)
82.  {
83.  }
84.
85.  /*************************************************************************************
86.  * 函数名称: SendAckPack
87.  * 函数功能: 发送响应包
88.  * 输入参数: moduleId，secondId，ackMsg
89.  * 输出参数: void
90.  * 返 回 值: void
91.  * 创建日期: 2021 年 03 月 01 日
92.  * 注    意:
93.  *************************************************************************************/
94.  void SendAckPack(u8 moduleId, u8 secondId, u8 ackMsg)
95.  {
96.    StructPackType pt;
97.
98.    pt.packModuleId = MODULE_SYS;
99.    pt.packSecondId = DAT_CMD_ACK;
100.   pt.arrData[0] = moduleId;
101.   pt.arrData[1] = secondId;
102.   pt.arrData[2] = ackMsg;
103.   pt.arrData[3] = 0;
104.   pt.arrData[4] = 0;
105.   pt.arrData[5] = 0;
106.
107.   SendPackToHost(&pt);//调用打包函数来打包数据，并将打包好的数据包发送到主机
108. }
109.
110. /*************************************************************************************
111. * 函数名称: SendWave1ToHost
112. * 函数功能: 发送 wave1 到主机，一次性发送 5 个点
113. * 输入参数: pWaveData-wave1 波形数据，0-255
114. * 输出参数: void
115. * 返 回 值: void
116. * 创建日期: 2021 年 03 月 01 日
117. * 注    意:
118. *************************************************************************************/
119. void  SendWave1ToHost(u8* pWaveData)
120. {
121.   StructPackType  pt;
122.
123.   pt.packModuleId = MODULE_WAVE1;
```

```
124.   pt.packSecondId = DAT_WAVE1_WDATA;
125.   pt.arrData[0] = pWaveData[0];
126.   pt.arrData[1] = pWaveData[1];
127.   pt.arrData[2] = pWaveData[2];
128.   pt.arrData[3] = pWaveData[3];
129.   pt.arrData[4] = pWaveData[4];
130.   pt.arrData[5] = 0;          //保留
131.
132.   SendPackToHost(&pt);        //调用打包函数来打包数据，并将打包好的数据包发送到主机
133. }
134.
135. /*********************************************************************************
136. * 函数名称: SendWave2ToHost
137. * 函数功能: 发送 wave2 到主机，一次性发送 5 个点
138. * 输入参数: pWaveData-wave2 波形数据，0-255
139. * 输出参数: void
140. * 返 回 值: void
141. * 创建日期: 2021 年 03 月 01 日
142. * 注    意:
143. *********************************************************************************/
144. void  SendWave2ToHost(u8* pWaveData)
145. {
146.   StructPackType  pt;
147.
148.   pt.packModuleId = MODULE_WAVE2;
149.   pt.packSecondId = DAT_WAVE2_WDATA;
150.   pt.arrData[0] = pWaveData[0];
151.   pt.arrData[1] = pWaveData[1];
152.   pt.arrData[2] = pWaveData[2];
153.   pt.arrData[3] = pWaveData[3];
154.   pt.arrData[4] = pWaveData[4];
155.   pt.arrData[5] = 0;          //保留
156.
157.   SendPackToHost(&pt);        //调用打包函数来打包数据，并把打包好的数据包发送到主机
158. }
159.
```

步骤 6：完善 App.c 文件

将程序清单 14-7 中的代码输入 App.c 文件中。该文件与第 10 章中的 App.c 文件类似，这里只解释第 110 至 154 行代码：每执行一次 Proc1SecTask 函数，变量 s_iCnt 循环递增计数，计数范围为 1～10，计数到 5 时，通过 ProcHostCmd 函数处理生成 wave1 的命令（生成 wave1 命令包格式参见图 14-3）；计数到 10 时，通过 ProcHostCmd 函数处理生成 wave2 的命令（生成 wave2 命令包格式参见图 14-4）。

程序清单 14-7

```
1.   /*********************************************************************************
2.   * 模块名称: App.c
3.   * 摘    要: 测试模拟从机命令接收和数据发送
4.   * 当前版本: 1.0.0
5.   * 作    者: Leyutek(COPYRIGHT 2018 - 2021 Leyutek. All rights reserved.)
6.   * 完成日期: 2021 年 03 月 01 日
```

```
7.  * 内      容:
8.  * 注      意:
9.  ********************************************************************************
10. * 取代版本:
11. * 作      者:
12. * 完成日期:
13. * 修改内容:
14. * 修改文件:
15. *******************************************************************************/
16.
17. /******************************************************************************
18. *                                 包含头文件
19. *******************************************************************************/
20. #include <stdio.h>
21. #include <windows.h>
22. #pragma comment(lib, "winmm.lib")    //导入 winmm.lib 多媒体库
23.
24. #include "DataType.h"
25. #include "PackUnpack.h"
26. #include "ProcHostCmd.h"
27. #include "SendDataToHost.h"
28.
29. /******************************************************************************
30. *                                  宏定义
31. *******************************************************************************/
32.
33. /******************************************************************************
34. *                               枚举结构体定义
35. *******************************************************************************/
36.
37. /******************************************************************************
38. *                                 内部变量
39. *******************************************************************************/
40.
41. /******************************************************************************
42. *                               内部函数声明
43. *******************************************************************************/
44. static  void  InitSoftware(void);      //初始化 Software
45.
46. static void __stdcall TimeProc(unsigned int uTimerID, unsigned int uMsg, unsigned long dwUser,
47.                                unsigned long dw1, unsigned long dw2);   //定时器回调函数
48.
49. static  void  Proc2msTask(void);       //声明一个 2ms 执行一次的函数
50. static  void  Proc1SecTask(void);      //声明一个 1s 执行一次的函数
51.
52. /******************************************************************************
53. *                               内部函数实现
54. *******************************************************************************/
```

```
55.  /*********************************************************************************************
56.  * 函数名称: InitSoftware
57.  * 函数功能: 所有的软件初始化函数都放在此函数中
58.  * 输入参数: void
59.  * 输出参数: void
60.  * 返 回 值: void
61.  * 创建日期: 2021 年 03 月 01 日
62.  * 注    意:
63.  *********************************************************************************************/
64.  static  void  InitSoftware(void)
65.  {
66.    InitPackUnpack();
67.    InitProcHostCmd();
68.    InitSendDataToHost();
69.  }
70.
71.  /*********************************************************************************************
72.  * 函数名称: TimeProc
73.  * 函数功能: 定时器回调函数
74.  * 输入参数: *
75.  * 输出参数: void
76.  * 返 回 值: void
77.  * 创建日期: 2021 年 03 月 01 日
78.  * 注    意:
79.  *********************************************************************************************/
80.  static void __stdcall TimeProc(unsigned int uTimerID, unsigned int uMsg, unsigned long dwUser,
81.                                 unsigned long dw1, unsigned long dw2)
82.  {
83.    Proc2msTask();
84.  }
85.
86.  /*********************************************************************************************
87.  * 函数名称: Proc2msTask
88.  * 函数功能: 处理 2ms 任务
89.  * 输入参数: void
90.  * 输出参数: void
91.  * 返 回 值: void
92.  * 创建日期: 2021 年 03 月 01 日
93.  * 注    意:
94.  *********************************************************************************************/
95.  static  void  Proc2msTask(void)
96.  {
97.    static  i16 s_iCnt500 = 0;
98.
99.    if(s_iCnt500 >=499)
100.   {
101.     Proc1SecTask();
102.     s_iCnt500 = 0;
```

```
103.   }
104.   else
105.   {
106.     s_iCnt500++;
107.   }
108. }
109.
110. /*********************************************************************************************
111. * 函数名称: Proc1SecTask
112. * 函数功能: 处理 1s 任务
113. * 输入参数: void
114. * 输出参数: void
115. * 返 回 值: void
116. * 创建日期: 2021 年 03 月 01 日
117. * 注    意:
118. *********************************************************************************************/
119. static  void  Proc1SecTask(void)
120. {
121.   static i16 s_iCnt = 0;
122.
123.   s_iCnt++;
124.
125.   if(s_iCnt == 5)
126.   {
127.     printf("\n");
128.     ProcHostCmd(0x70);
129.     ProcHostCmd(0x81);
130.     ProcHostCmd(0x80);
131.     ProcHostCmd(0x80);
132.     ProcHostCmd(0x80);
133.     ProcHostCmd(0x80);
134.     ProcHostCmd(0x80);
135.     ProcHostCmd(0x80);
136.     ProcHostCmd(0x80);
137.     ProcHostCmd(0xF1);
138.   }
139.   else if(s_iCnt == 10)
140.   {
141.     printf("\n");
142.     ProcHostCmd(0x71);
143.     ProcHostCmd(0x81);
144.     ProcHostCmd(0x80);
145.     ProcHostCmd(0x80);
146.     ProcHostCmd(0x80);
147.     ProcHostCmd(0x80);
148.     ProcHostCmd(0x80);
149.     ProcHostCmd(0x80);
150.     ProcHostCmd(0x80);
```

```
151.     ProcHostCmd(0xF2);
152.     s_iCnt = 0;
153.   }
154. }
155.
156. /*********************************************************************************************
157. *                                        API 函数实现
158. *********************************************************************************************/
159. /*********************************************************************************************
160. * 函数名称: main
161. * 函数功能: 主函数
162. * 输入参数: void
163. * 输出参数: void
164. * 返 回 值: void
165. * 创建日期: 2021 年 03 月 01 日
166. * 注    意:
167. *********************************************************************************************/
168. void  main(void)
169. {
170.   InitSoftware();                          //初始化软件
171.
172.   //timeSetEvent 函数说明
173.   //MMRESULT timeSetEvent(
174.   //  UINT             uDelay,              //以 ms 指定事件的周期
175.   //  UINT             uResolution,         //以 ms 指定延时的精度，数值越小定时器事件分辨率越高，
                                                 默认值为 1ms
176.   //  LPTIMECALLBACK lpTimeProc,            //指向一个回调函数，即单次事件或周期性事件触发时调用
                                                 的函数
177.   //  DWORD_PTR        dwUser,              //存放用户提供的回调数据
178.   //  UINT             fuEvent                     //指定定时器事件类型，TIME_ONESHOT-单次触发，
                                                 TIME_ PERIODIC-周期性触发
179.   //);
180.
181.   //用户定时器设定，定时器精度为 1ms，每隔 2ms 触发一次定时器，并执行回调函数 TimeProc
182.   timeSetEvent(2, 1, TimeProc, 0, TIME_PERIODIC);
183.
184.   while(1)
185.   {
186.   }
187. }
```

步骤 7：项目编译和运行

最后，按 F5 键编译并运行程序，在弹出的控制台窗口中，可以看到如图 14-9 所示的运行结果，即从机接收到主机发送的生成 wave1 的命令（0x70, 0x81, 0x80, 0x80, 0x80, 0x80, 0x80, 0x80, 0x80, 0xF1）后，向主机发送响应包（0x01, 0x84, 0x84, 0xF0, 0x80, 0x80, 0x80, 0x80, 0x80, 0xF9），然后不断地向主机发送数据包（0x70, 0x80, 0x82, 0x91, 0x91, 0x91, 0x91, 0x91, 0x80, 0xC7）；同样，当从机接收到主机发送的生成 wave2 命令（0x71, 0x81, 0x80, 0x80, 0x80, 0x80,

0x80, 0x80, 0x80, 0xF2）后，向主机发送响应包（0x01, 0x84, 0x84, 0xF1, 0x80, 0x80, 0x80, 0x80, 0x80, 0xFA），然后不断地向主机发送数据包（0x71, 0x80, 0x82, 0xA2, 0xA2, 0xA2, 0xA2, 0xA2, 0x80, 0x9D），说明实验成功。

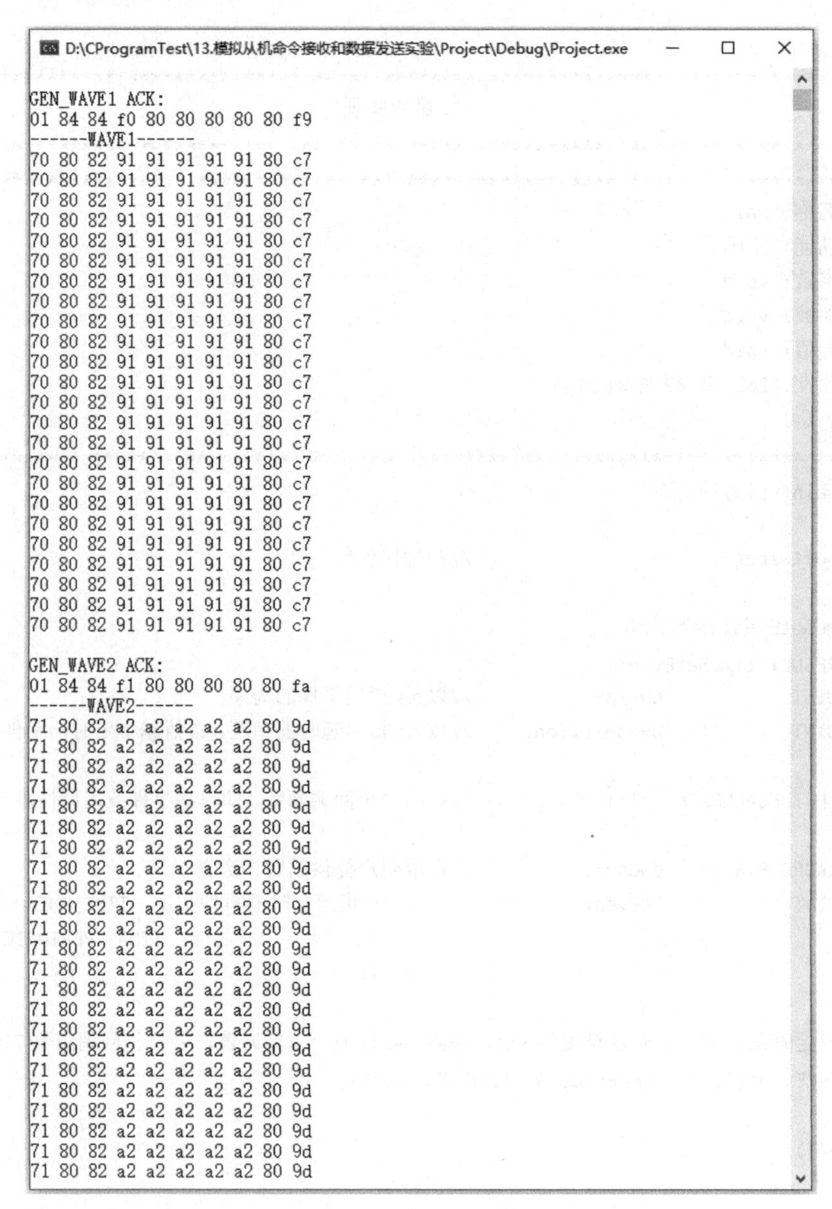

图 14-9　本章项目运行结果

本 章 任 务

使用智能小车作为从机模块，模块 ID 为 0x13（MODULE_CAR）。通过主机（如手机）查询智能小车的空间信息，这里规定通过主机查询空间信息命令包的二级 ID 为 0x80（CMD_CAR_QUERY_POS），该命令包的数据 1～数据 6 均默认为 0x00。主机向从机发送查询智能小车的空间信息命令包之后，从机应向主机发送一个命令应答数据包，命令应答数据包属于系统模块，其模块 ID 为 0x01（MODULE_SYS），命令应答数据包的二级 ID 为 0x04（DAT_CMD_ACK）。该数据包的数据 1 和数据 2 分别为查询智能小车空间信息命令包的模块 ID 和二级 ID，数据 3 为应答消息（假设为 0x00，表示接收命令成功），数据 4～数据 6 均默认为 0x00。发送完命令应答数据包之后，从机再向主机发送空间信息数据包，这里规定空间信息数据包的二级 ID 为 0x03（DAT_CAR_POS）。该数据包的数据 1～数据 4 分别为智能小车距离前方、后方、左侧和右侧障碍物的距离（单位为 cm），数据 5 和数据 6 默认为 0x00。在发送位置信息时，假设数据 1～数据 4 分别为 25、70、30 和 30，数据 5 和数据 6 均为 0x00。

参考本章实验，主机每 5s 发送一次查询智能小车空间信息命令包，从机接收到该命令包之后，向主机发送命令应答数据包，接着向主机发送空间信息数据包。发送命令应答数据包和空间信息数据包通过 printf 函数打印到控制台窗口中。

本 章 习 题

1．简述命令从主机到从机的通信过程。

2．简述数据从从机到主机的通信过程。

3．本章实验中的 ProcHostCmd 模块包含哪些 API 函数？简述每个 API 函数的功能。

4．本章实验中的 SendDataToHost 模块包含哪些 API 函数？简述每个 API 函数的功能。

第 15 章　模拟主机命令发送与数据接收

本章实验通过模拟的方式实现主机命令发送与数据接收，命令发送是指主机对命令进行打包和发送，数据接收是指主机对接收到的数据进行接收、解包和处理。最后，在 App.c 文件中对主机命令发送与数据接收进行测试验证。

15.1　实验内容

使用第 13 章中的 PCT 通信协议，模拟主机命令发送和数据接收。将主机假设为计算机，将从机假设为单片机，如图 15-1 所示，模拟主机命令发送与数据接收的具体流程如下：①主机准备生成波形的命令，并对该命令进行打包处理，然后将打包好的生成波形命令包发送到从机；②主机处理接收到的从机数据，当接收到一个完整数据包时，对其进行解包处理，最后对解包后的波形数据进行处理，包括将其显示到屏幕上。

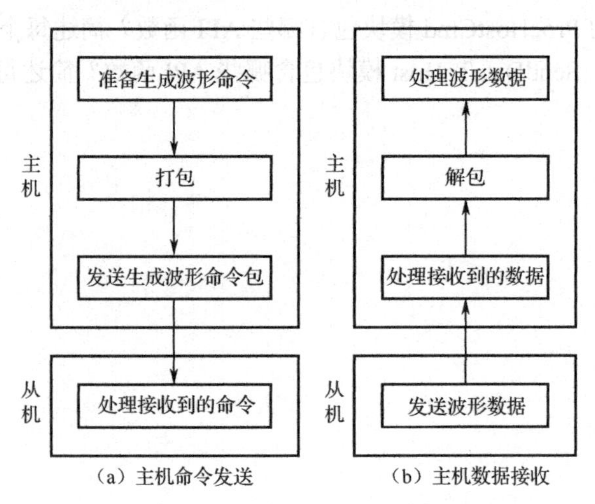

图 15-1　主机命令发送和数据接收

15.2　实验原理

15.2.1　主机命令发送流程说明

如图 15-2 所示，主机命令发送流程如下：①在主函数中启动一个 1s 定时器，该定时器实质上是间接通过 2ms 多媒体定时器启动的；②在 Proc1SecTask 函数中判断 1s 计数是否溢出，如果未产生溢出，则继续判断；③如果产生溢出，则发送命令包到从机。

15.2.2　主机数据接收流程说明

如图 15-3 所示，主机数据接收流程如下：①启动 20ms 定时器；②判断 20ms 计数是否溢出，如果未产生溢出，则继续判断；③如果产生溢出，则处理从机数据，并对从机数据进行解包；④判断解包结果，如果接收到完整包，则从解包结果中获取解包后的数据；⑤判断是否收到 wave1 数据，如果收到该数据，则打印 wave1 数据；⑥如果未收到 wave1 数据，则继续判断是否收到 wave2 数据，如果收到该数据，则打印 wave2 数据；⑦如果既未收到 wave1 数据，也未收到 wave2 数据，则判断 20ms 计数是否溢出。

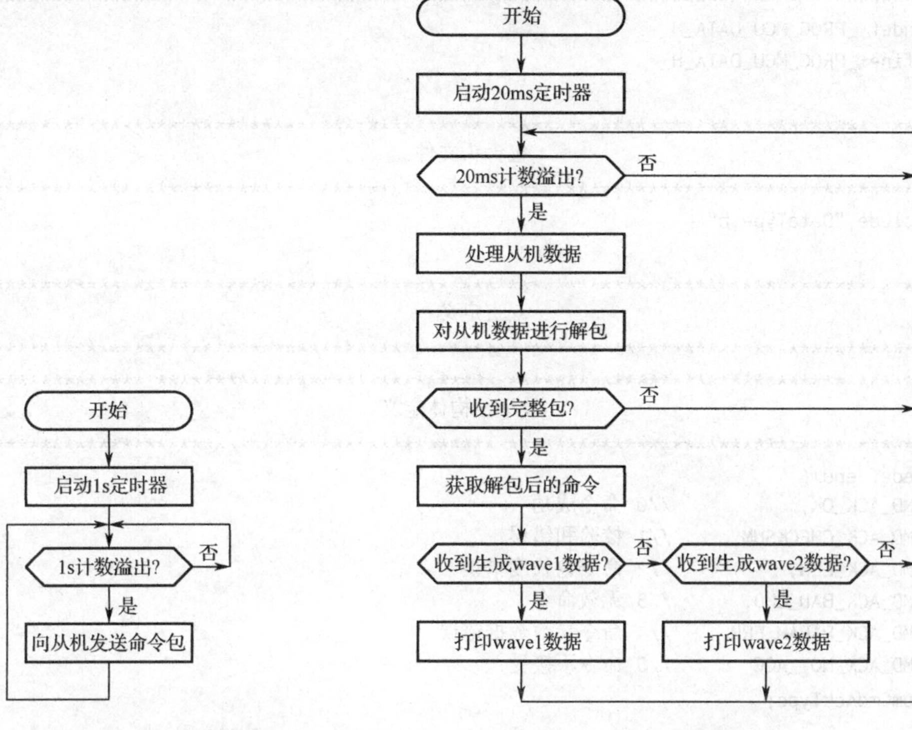

图 15-2 主机命令发送流程图

图 15-3 主机数据接收流程图

15.3 实验步骤

步骤 1：复制原始项目

首先，将本书配套资料包的“03.例程资料\Material\14.模拟主机命令发送与数据接收实验”文件夹复制到 CProgramTest 文件夹中，然后，双击运行“D:\CProgramTest\14.模拟主机命令发送与数据接收实验\Project”文件夹中的 Project.sln 文件。

步骤 2：完善 ProcMCUData.h 文件

将程序清单 15-1 中的代码输入 ProcMCUData.h 文件中。该文件与 ProcHostCmd.h 类似，这里不作解释。

程序清单 15-1

```
1.  /*********************************************************************************
2.  * 模块名称: ProcMCUData.h
3.  * 摘    要: 处理接收到的从机数据
4.  * 当前版本: 1.0.0
5.  * 作    者: Leyutek(COPYRIGHT 2018 - 2021 Leyutek. All rights reserved.)
6.  * 完成日期: 2021 年 03 月 01 日
7.  * 内    容:
8.  * 注    意:
9.  **********************************************************************************
10. * 取代版本:
```

```
11.  * 作    者:
12.  * 完成日期:
13.  * 修改内容:
14.  * 修改文件:
15.  *********************************************************************************/
16.  #ifndef _PROC_MCU_DATA_H_
17.  #define _PROC_MCU_DATA_H_
18.
19.  /*********************************************************************************
20.  *                                   包含头文件
21.  *********************************************************************************/
22.  #include "DataType.h"
23.
24.  /*********************************************************************************
25.  *                                     宏定义
26.  *********************************************************************************/
27.  /*********************************************************************************
28.  *                                  枚举结构体定义
29.  *********************************************************************************/
30.  typedef enum{
31.    CMD_ACK_OK,             //0 命令成功
32.    CMD_ACK_CHECKSUM,       //1 校验和错误
33.    CMD_ACK_LEN,            //2 命令包长度错误
34.    CMD_ACK_BAD_CMD,        //3 无效命令
35.    CMD_ACK_PARAM_ERR,      //4 命令参数数据错误
36.    CMD_ACK_NOT_ACC         //5 命令不接受
37.  }EnumCmdAckType;
38.
39.
40.  /*********************************************************************************
41.  *                                  API 函数声明
42.  *********************************************************************************/
43.  void  InitProcMCUData(void);        //初始化 ProcMCUData 模块
44.  void  ProcMCUData(u8 recData);      //处理来自从机的数据
45.
46.  #endif
```

步骤 3：完善 ProcMCUData.c 文件

将程序清单 15-2 中的代码输入 ProcMCUData.c 文件中。下面按照顺序对部分语句进行解释。

（1）第 44 至 76 行代码：一般主机在接收到从机发送的数据时，会对其进行处理，例如，将其显示到软件界面或存储到数据库等，本章只是为了模拟主机数据接收，因此，OnRecWave1 函数在收到从机发送的数据后，将其通过 printf 函数输出到控制台。OnRecWave2 与 OnRecWave1 的功能类似。

（2）第 95 至 129 行代码：ProcMCUData 函数用于处理从机发送到主机的未解包的数据，该函数每隔一定的时间处理一个接收到的数据，将 UnPackData 函数的返回值作为 while 循环语句的表达式，当该函数的返回值为真时，即接收到完整数据包时，通过 GetUnPackRslt 函数获取解包后的结果，并将其赋值给 pack，根据 pack 的 packModuleId 成员可以判断接收到的数据是 wave1 还是 wave2 的波形数据。如果接收到的是 wave1 波形数据，则执行 OnRecWave1 函数；如果接收到的是 wave2 波形数据，则执行 OnRecWave2 函数。

程序清单 15-2

```
1.  /*********************************************************************************
2.  * 模块名称: ProcMCUData.c
3.  * 摘    要: 处理接收到的从机数据
4.  * 当前版本: 1.0.0
5.  * 作    者: Leyutek(COPYRIGHT 2018 - 2021 Leyutek. All rights reserved.)
6.  * 完成日期: 2021 年 03 月 01 日
7.  * 内    容:
8.  * 注    意:
9.  **********************************************************************************
10. * 取代版本:
11. * 作    者:
12. * 完成日期:
13. * 修改内容:
14. * 修改文件:
15. *********************************************************************************/
16.
17. /*********************************************************************************
18. *                                   包含头文件
19. *********************************************************************************/
20. #include "ProcMCUData.h"
21. #include "PackUnpack.h"
22. #include "SendCmdToMCU.h"
23.
24. #include <stdio.h>
25.
26. /*********************************************************************************
27. *                                   宏定义
28. *********************************************************************************/
29.
30. /*********************************************************************************
31. *                                   内部变量
32. *********************************************************************************/
33.
34. /*********************************************************************************
35. *                                  内部函数声明
36. *********************************************************************************/
37. static  u8 OnRecWave1(u8* pMsg);       //接收 wave1 的响应函数声明
38. static  u8 OnRecWave2(u8* pMsg);       //接收 wave2 的响应函数声明
39.
40. /*********************************************************************************
41. *                                   内部函数实现
42. *********************************************************************************/
43.
44. /*********************************************************************************
45. * 函数名称: OnRecWave1
46. * 函数功能: 生成 wave1
47. * 输入参数: pMsg
48. * 输出参数: void
49. * 返 回 值: 响应结果
50. * 创建日期: 2021 年 03 月 01 日
51. * 注    意:
```

```
52. ****************************************************************************************/
53. static  u8 OnRecWave1(u8* pMsg)
54. {
55.   printf("Print rec wave1 data:0x%02x, 0x%02x, 0x%02x, 0x%02x, 0x%02x, 0x%02x\n", pMsg[0],
56.     pMsg[1], pMsg[2], pMsg[3], pMsg[4], pMsg[5]);
57.
58.   return(CMD_ACK_OK);
59. }
60.
61. /****************************************************************************************
62. * 函数名称: OnRecWave2
63. * 函数功能: 生成 wave2
64. * 输入参数: pMsg
65. * 输出参数: void
66. * 返 回 值: 响应结果
67. * 创建日期: 2021 年 03 月 01 日
68. * 注    意:
69. ****************************************************************************************/
70. static  u8 OnRecWave2(u8* pMsg)
71. {
72.   printf("Print rec wave2 data:0x%02x, 0x%02x, 0x%02x, 0x%02x, 0x%02x, 0x%02x\n", pMsg[0],
73.     pMsg[1], pMsg[2], pMsg[3], pMsg[4], pMsg[5]);
74.
75.   return(CMD_ACK_OK);
76. }
77.
78. /****************************************************************************************
79. *                                    API 函数实现
80. ****************************************************************************************/
81.
82. /****************************************************************************************
83. * 函数名称: InitProcMCUData
84. * 函数功能: 初始化该模块
85. * 输入参数: void
86. * 输出参数: void
87. * 返 回 值: void
88. * 创建日期: 2021 年 03 月 01 日
89. * 注    意:
90. ****************************************************************************************/
91. void  InitProcMCUData(void)
92. {
93. }
94.
95. /****************************************************************************************
96. * 函数名称: ProcMCUData
97. * 函数功能: 处理从机发送来的数据
98. * 输入参数: recData
99. * 输出参数: void
100. * 返 回 值: void
101. * 创建日期: 2021 年 03 月 01 日
102. * 注    意:
103. ****************************************************************************************/
```

```
104. void ProcMCUData(u8 recData)
105. {
106.   StructPackType pack;
107.
108.   while(UnPackData(recData))
109.   {
110.     pack = GetUnPackRslt();                   //获取解包结果
111.     switch(pack.packModuleId)
112.     {
113.       case MODULE_WAVE1:
114.         if(DAT_WAVE1_WDATA == pack.packSecondId)
115.         {
116.           OnRecWave1(pack.arrData);
117.         }
118.         break;
119.       case MODULE_WAVE2:
120.         if(DAT_WAVE2_WDATA == pack.packSecondId)
121.         {
122.           OnRecWave2(pack.arrData);
123.         }
124.         break;
125.       default:
126.         break;
127.     }
128.   }
129. }
```

步骤 4：完善 SendCmdToMCU.h 文件

将程序清单 15-3 中的代码输入 SendCmdToMCU.h 文件中。在 API 函数声明区，声明 3 个函数，分别是初始化 SendCmdToMCU 模块函数 InitSendCmdToMCU、发送生成 wave1 命令到从机函数 SendGenWave1CmdToMCU 和发送生成 wave2 命令到从机函数 SendGenWave2CmdToMCU。

程序清单 15-3

```
1.  /*********************************************************************************
2.  * 模块名称: SendCmdToMCU.h
3.  * 摘    要: 发送命令到从机
4.  * 当前版本: 1.0.0
5.  * 作    者: Leyutek(COPYRIGHT 2018 - 2021 Leyutek. All rights reserved.)
6.  * 完成日期: 2021 年 03 月 01 日
7.  * 内    容:
8.  * 注    意:
9.  **********************************************************************************
10. * 取代版本:
11. * 作    者:
12. * 完成日期:
13. * 修改内容:
14. * 修改文件:
15. *********************************************************************************/
16. #ifndef _SEND_CMD_TO_MCU_H_
17. #define _SEND_CMD_TO_MCU_H_
18.
```

```
19. /****************************************************************************************
20. *                                  包含头文件
21. ****************************************************************************************/
22. #include "DataType.h"
23.
24. /****************************************************************************************
25. *                                   宏定义
26. ****************************************************************************************/
27.
28. /****************************************************************************************
29. *                                枚举结构体定义
30. ****************************************************************************************/
31.
32.
33. /****************************************************************************************
34. *                                 API 函数声明
35. ****************************************************************************************/
36. void  InitSendCmdToMCU(void);          //初始化 SendCmdToMCU 模块
37.
38. void  SendGenWave1CmdToMCU(void);      //发送生成 wave1 命令到从机
39. void  SendGenWave2CmdToMCU(void);      //发送生成 wave2 命令到从机
40.
41. #endif
```

步骤 5：完善 SendCmdToMCU.c 文件

将程序清单 15-4 中的代码输入 SendCmdToMCU.c 文件中。下面按照顺序对部分语句进行解释。

（1）第 45 至 68 行代码：SendCmdToMCU 模块中的 SendPackToMCU 函数功能与 SendDataToHost 模块中的 SendPackToHost 函数功能类似，这里不作解释。

（2）第 87 至 135 行代码：SendGenWave1CmdToMCU 函数的功能是发送生成 wave1 命令到从机。SendGenWave2CmdToMCU 与 SendGenWave1CmdToMCU 的功能类似。

程序清单 15-4

```
1.  /****************************************************************************************
2.  * 模块名称: SendCmdToMCU.c
3.  * 摘    要: 发送命令到从机
4.  * 当前版本: 1.0.0
5.  * 作    者: Leyutek(COPYRIGHT 2018 - 2021 Leyutek. All rights reserved.)
6.  * 完成日期: 2021 年 03 月 01 日
7.  * 内    容:
8.  * 注    意:
9.  *****************************************************************************************
10. * 取代版本:
11. * 作    者:
12. * 完成日期:
13. * 修改内容:
14. * 修改文件:
15. ****************************************************************************************/
16.
17. /****************************************************************************************
18. *                                  包含头文件
```

```
19.  *****************************************************************************/
20.  #include "SendCmdToMCU.h"
21.  #include "PackUnpack.h"
22.
23.  #include <stdio.h>
24.
25.  /******************************************************************************
26.  *                                   宏定义
27.  *****************************************************************************/
28.
29.  /******************************************************************************
30.  *                                 枚举结构体定义
31.  *****************************************************************************/
32.
33.  /******************************************************************************
34.  *                                  内部变量
35.  *****************************************************************************/
36.
37.  /******************************************************************************
38.  *                                 内部函数声明
39.  *****************************************************************************/
40.  static  void  SendPackToMCU(StructPackType* packSent);     //发送 Pack 到从机
41.
42.  /******************************************************************************
43.  *                                 内部函数实现
44.  *****************************************************************************/
45.  /******************************************************************************
46.  * 函数名称: SendPackToMCU
47.  * 函数功能: 发送 Pack 到从机
48.  * 输入参数: pPackSent-包类型结构体指针
49.  * 输出参数: void
50.  * 返 回 值: void
51.  * 创建日期: 2021 年 03 月 01 日
52.  * 注    意:
53.  *****************************************************************************/
54.  static  void  SendPackToMCU(StructPackType* pPackSent)
55.  {
56.    u8  valid;
57.    i16 i = 0;
58.
59.    valid = PackData(pPackSent);                       //调用打包函数来打包数据
60.
61.    if(1 == valid)
62.    {
63.      for(i = 0; i < 10; i++)
64.      {
65.        printf("%02x ", *((u8 *)pPackSent + i)); //将打包好的数据发送出去，这里使用 printf
                                                       进行模拟
66.      }
67.    }
68.  }
69.
```

```
70. /*********************************************************************************
71. *                                        API 函数实现
72. *********************************************************************************/
73.
74. /*********************************************************************************
75. * 函数名称: InitSendCmdToMCU
76. * 函数功能: 初始化 SendCmdToMCU 模块
77. * 输入参数: void
78. * 输出参数: void
79. * 返 回 值: void
80. * 创建日期: 2021 年 03 月 01 日
81. * 注    意:
82. *********************************************************************************/
83. void  InitSendCmdToMCU(void)
84. {
85. }
86.
87. /*********************************************************************************
88. * 函数名称: SendGenWave1CmdToMCU
89. * 函数功能: 发送生成 wave1 命令到从机
90. * 输入参数: void
91. * 输出参数: void
92. * 返 回 值: void
93. * 创建日期: 2021 年 03 月 01 日
94. * 注    意:
95. *********************************************************************************/
96. void  SendGenWave1CmdToMCU(void)
97. {
98.   StructPackType  pt;
99.
100.  pt.packModuleId = MODULE_WAVE1;
101.  pt.packSecondId = CMD_GEN_WAVE1;
102.  pt.arrData[0] = 0;
103.  pt.arrData[1] = 0;
104.  pt.arrData[2] = 0;
105.  pt.arrData[3] = 0;
106.  pt.arrData[4] = 0;
107.  pt.arrData[5] = 0;
108.
109.  SendPackToMCU(&pt);       //调用打包函数来打包数据，并将数据发送到从机
110. }
111.
112. /*********************************************************************************
113. * 函数名称: SendGenWave2CmdToMCU
114. * 函数功能: 发送生成 wave2 命令到从机
115. * 输入参数: void
116. * 输出参数: void
117. * 返 回 值: void
118. * 创建日期: 2021 年 03 月 01 日
119. * 注    意:
120. *********************************************************************************/
121. void  SendGenWave2CmdToMCU(void)
```

```
122. {
123.   StructPackType  pt;
124.
125.   pt.packModuleId = MODULE_WAVE2;
126.   pt.packSecondId = CMD_GEN_WAVE2;
127.   pt.arrData[0] = 0;
128.   pt.arrData[1] = 0;
129.   pt.arrData[2] = 0;
130.   pt.arrData[3] = 0;
131.   pt.arrData[4] = 0;
132.   pt.arrData[5] = 0;
133.
134.   SendPackToMCU(&pt);           //调用打包函数来打包数据，并将数据发送到从机
135. }
```

步骤 6：完善 App.c 文件

将程序清单 15-5 中的代码输入 App.c 文件中。下面按照顺序对部分语句进行解释。

（1）第 86 至 119 行代码：每执行一次 Proc2msTask 函数，变量 s_iCnt10 循环递增计数，计数范围为 0～9，当计数到 9 时，执行 Proc20msTask 函数。Proc2msTask 函数每 2ms 执行一次，因此，Proc20msTask 函数每 20ms 执行一次。

（2）第 121 至 150 行代码：s_arrRecMCUData 数组中有 20 个数据，分别是 wave1 和 wave2 数据包，这两个数据包中均包含 10 字节数据。每执行一次 Proc20msTask 函数，变量 s_iCnt 循环递增计数，计数范围为 0～19，ProcMCUData 函数每 20ms 读取处理一个 s_arrRecMCUData 数组中的数据，因此，每 400ms 就可以处理完一次 wave1 和 wave2 数据包。

程序清单 15-5

```
1.  /*********************************************************************************
2.  * 模块名称: App.c
3.  * 摘    要: 测试模拟主机命令发送和数据接收
4.  * 当前版本: 1.0.0
5.  * 作    者: Leyutek(COPYRIGHT 2018 - 2021 Leyutek. All rights reserved.)
6.  * 完成日期: 2021 年 03 月 01 日
7.  * 内    容:
8.  * 注    意:
9.  **********************************************************************************
10. * 取代版本:
11. * 作    者:
12. * 完成日期:
13. * 修改内容:
14. * 修改文件:
15. *********************************************************************************/
16.
17. /*********************************************************************************
18. *                                包含头文件
19. *********************************************************************************/
20. #include <stdio.h>
21. #include <windows.h>
22. #pragma comment(lib, "winmm.lib")     //导入 winmm.lib 多媒体库
23.
24. #include "PackUnpack.h"
```

```
25.  #include "ProcMCUData.h"
26.  #include "SendCmdToMCU.h"
27.
28.  /*********************************************************************************
29.  *                                   宏定义
30.  *********************************************************************************/
31.
32.  /*********************************************************************************
33.  *                                枚举结构体定义
34.  *********************************************************************************/
35.
36.  /*********************************************************************************
37.  *                                  内部变量
38.  *********************************************************************************/
39.
40.  /*********************************************************************************
41.  *                                内部函数声明
42.  *********************************************************************************/
43.  static  void  InitSoftware(void);      //初始化 Software
44.
45.  static void __stdcall TimeProc(unsigned int uTimerID, unsigned int uMsg, unsigned long dwUser,
46.                               unsigned long dw1, unsigned long dw2);
47.
48.  static  void  Proc2msTask(void);          //声明一个 2ms 执行一次的函数
49.  static  void  Proc20msTask(void);         //声明一个 20ms 执行一次的函数
50.  static  void  Proc1SecTask(void);         //声明一个 1s 执行一次的函数
51.
52.  /*********************************************************************************
53.  *                                内部函数实现
54.  *********************************************************************************/
55.  /*********************************************************************************
56.  * 函数名称: InitSoftware
57.  * 函数功能: 所有的软件初始化函数都放在此函数中
58.  * 输入参数: void
59.  * 输出参数: void
60.  * 返 回 值: void
61.  * 创建日期: 2021 年 03 月 01 日
62.  * 注    意:
63.  *********************************************************************************/
64.  static  void  InitSoftware(void)
65.  {
66.    InitPackUnpack();
67.    InitProcMCUData();
68.    InitSendCmdToMCU();
69.  }
70.
71.  /*********************************************************************************
72.  * 函数名称: TimeProc
73.  * 函数功能: 定时器回调函数
74.  * 输入参数: *
75.  * 输出参数: void
76.  * 返 回 值: void
```

```
77. * 创建日期: 2021 年 03 月 01 日
78. * 注    意:
79. *********************************************************************************/
80. static void __stdcall TimeProc (unsigned int uTimerID, unsigned int uMsg, unsigned long dwUser,
81.                               unsigned long dw1, unsigned long dw2)
82. {
83.   Proc2msTask();
84. }
85.
86. /*********************************************************************************
87. * 函数名称: Proc2msTask
88. * 函数功能: 处理 2ms 任务
89. * 输入参数: void
90. * 输出参数: void
91. * 返 回 值: void
92. * 创建日期: 2021 年 03 月 01 日
93. * 注    意:
94. *********************************************************************************/
95. static  void  Proc2msTask(void)
96. {
97.   static  i16 s_iCnt500 = 0;
98.   static  i16 s_iCnt10  = 0;
99.
100.  if(s_iCnt10 >= 9)
101.  {
102.    Proc20msTask();
103.    s_iCnt10 = 0;
104.  }
105.  else
106.  {
107.    s_iCnt10++;
108.  }
109.
110.  if(s_iCnt500 >= 499)
111.  {
112.    Proc1SecTask();
113.    s_iCnt500 = 0;
114.  }
115.  else
116.  {
117.    s_iCnt500++;
118.  }
119. }
120.
121. /*********************************************************************************
122. * 函数名称: Proc20msTask
123. * 函数功能: 处理 20ms 任务
124. * 输入参数: void
125. * 输出参数: void
126. * 返 回 值: void
127. * 创建日期: 2021 年 03 月 01 日
128. * 注    意:
```

```
129. **************************************************************************************/
130. static  void  Proc20msTask(void)
131. {
132.   static  i16 s_iCnt = 0;
133.
134.   //模拟接收到的 MCU 数据，包含 2 个波形数据包
135.   static  u8  s_arrRecMCUData[20] = {0x70, 0xa0, 0x82, 0x92, 0xb4, 0xd6, 0xf8, 0x90, 0x80,
136.                             0xb6, 0x71, 0xa0, 0x82, 0x91, 0xb3, 0xd5, 0xf7, 0x99, 0x80, 0xbc};
137.
138.   //每隔 20ms 处理一次从机数据，即对接收到的数据进行解包处理
139.   ProcMCUData(s_arrRecMCUData[s_iCnt]);
140.
141.   //以下的循环计数是为了循环处理 s_arrRecMCUData 中的数据
142.   if(s_iCnt >= 19)
143.   {
144.     s_iCnt = 0;
145.   }
146.   else
147.   {
148.     s_iCnt++;
149.   }
150. }
151.
152. /**************************************************************************************
153. * 函数名称: Proc1SecTask
154. * 函数功能: 处理 1s 任务
155. * 输入参数: void
156. * 输出参数: void
157. * 返 回 值: void
158. * 创建日期: 2021 年 03 月 01 日
159. * 注    意:
160. **************************************************************************************/
161. static  void  Proc1SecTask(void)
162. {
163.   printf("Print gen wave1 cmd:");
164.   SendGenWave1CmdToMCU();
165.   printf("\n");
166.
167.   printf("Print gen wave2 cmd:");
168.   SendGenWave2CmdToMCU();
169.   printf("\n");
170. }
171.
172. /**************************************************************************************
173. *                                API 函数实现
174. **************************************************************************************/
175. /**************************************************************************************
176. * 函数名称: main
177. * 函数功能: 主函数
178. * 输入参数: void
179. * 输出参数: void
180. * 返 回 值: void
```

```
181. * 创建日期: 2021 年 03 月 01 日
182. * 注      意:
183. **********************************************************************************/
184. void  main(void)
185. {
186.   InitSoftware();                              //初始化软件
187.
188.   //timeSetEvent 函数说明
189.   //MMRESULT timeSetEvent(
190.   //  UINT           uDelay,                    //以 ms 指定事件的周期
191.   //  UINT           uResolution,               //以 ms 指定延时的精度，数值越小定时器事件分辨率越高，
                                                      默认值为 1ms
192.   //  LPTIMECALLBACK lpTimeProc,                //指向一个回调函数，即单次事件或周期性事件触发时调用
                                                      的函数
193.   //  DWORD_PTR      dwUser,                    //存放用户提供的回调数据
194.   //  UINT           fuEvent                           //指定定时器事件类型，TIME_ONESHOT-单次触发，
                                                      TIME_ PERIODIC-周期性触发
195.   //);
196.
197.   //用户定时器设定，定时器精度为 1ms，每隔 2ms 触发一次定时器，并执行回调函数 TimeProc
198.   timeSetEvent(2, 1, TimeProc, 0, TIME_PERIODIC);
199.
200.   while(1)
201.   {
202.   }
203. }
```

步骤 7：项目编译和运行

最后，按 F5 键编译并运行程序，在弹出的控制台窗口中，可以看到如图 15-4 所示的运行结果，说明实验成功。

```
D:\CProgramTest\14.模拟主机命令发送和数据接收实验\Project\Debug\Project.exe
Print rec wave1 data:0x12, 0x34, 0x56, 0x78, 0x90, 0x00
Print rec wave2 data:0x11, 0x33, 0x55, 0x77, 0x99, 0x00
Print rec wave1 data:0x12, 0x34, 0x56, 0x78, 0x90, 0x00
Print rec wave2 data:0x11, 0x33, 0x55, 0x77, 0x99, 0x00
Print rec wave1 data:0x12, 0x34, 0x56, 0x78, 0x90, 0x00
Print gen wave1 cmd:70 81 80 80 80 80 80 80 80 f1
Print gen wave2 cmd:71 81 80 80 80 80 80 80 80 f2
Print rec wave2 data:0x11, 0x33, 0x55, 0x77, 0x99, 0x00
Print rec wave1 data:0x12, 0x34, 0x56, 0x78, 0x90, 0x00
Print rec wave2 data:0x11, 0x33, 0x55, 0x77, 0x99, 0x00
Print rec wave1 data:0x12, 0x34, 0x56, 0x78, 0x90, 0x00
Print rec wave2 data:0x11, 0x33, 0x55, 0x77, 0x99, 0x00
Print gen wave1 cmd:70 81 80 80 80 80 80 80 80 f1
Print gen wave2 cmd:71 81 80 80 80 80 80 80 80 f2
```

图 15-4　本章项目运行结果

本 章 任 务

使用智能小车作为从机模块，模块 ID 为 0x13（MODULE_CAR）。主机（如手机）通过向从机发送命令包，可以读取或中止读取智能小车的速度信息，这里规定读取速度信息命令包的二级 ID 为 0x90（CMD_CAR_QUERY_SPEED），中止读取速度信息命令包的二级 ID 为 0x91（CMD_CAR_STOP_QUERY），这两个命令包的数据 1～数据 6 均默认为 0x00。

参考本章实验，编写程序实现主机每隔 4s 向从机交替发送读取和中止读取速度信息命令，从机在收到读取或中止读取命令后，先向主机发送一个命令应答数据包，命令应答数据

包属于系统模块，其模块 ID 为 0x01（MODULE_SYS），命令应答数据包的二级 ID 为 0x04（DAT_CMD_ACK），该数据包的数据 1 和数据 2 分别为读取或中止读取速度信息命令包的模块 ID 和二级 ID，数据 3 为应答消息（假设为 0x00，表示接收命令成功），数据 4～数据 6 均默认为 0x00。

发送完读取速度信息命令应答数据包之后，从机每 200ms 向主机发送一次速度信息数据包，这里规定速度信息数据包的二级 ID 为 0x04（DAT_CAR_SPEED），该数据包的数据 1 为速度信息（单位为 cm/s），数据 2～数据 6 均默认为 0x00。发送速度信息时，规定速度从 25 开始，每发送一次递减 1，如果递减到 0，则继续从 25 开始递减。

发送完中止读取速度信息命令应答数据包之后，从机中止发送速度信息数据包，直到从机再次收到读取速度信息命令包，继续开始每 200ms 向主机发送速度信息数据包，且速度值从上一次中止时的速度值开始递减。

主机向从机发送的读取和中止读取速度信息命令包，以及从机向主机发送的应答数据包和速度信息数据包，均通过 printf 函数打印到控制台窗口中。

本 章 习 题

1．简述主机发送命令的详细流程。
2．命令主机接收数据的详细流程。
3．本章实验中的 ProcMCUData 模块包含哪些 API 函数？简述每个 API 函数的功能。
4．本章实验中的 SendCmdToMCU 模块包含哪些 API 函数？简述每个 API 函数的功能。

附录A　C语言软件设计规范（LY-STD001-2019）

该规范是由深圳市乐育科技有限公司于 2019 年发布的 C 语言软件设计规范，版本为LY-STD001-2019。该规范详细介绍了C语言的书写规范，包括排版、注释、命名规范等，给出了C文件模板和H文件模板，并对这两个模板进行了详细的说明。按照规范编写程序，可以使程序代码更加规范和高效，对代码的理解和维护起到至关重要的作用。

A.1　排版

（1）程序块采用缩进风格编写，缩进 2 个空格。对于由开发工具自动生成的代码可以不一致。

（2）须将 Tab 键设定为 2 个空格，以免用不同的编辑器阅读程序时，因 Tab 键所设置的空格数目不同而造成程序布局不整齐。对于由开发工具自动生成的代码可以不一致。

（3）相对独立的程序块之间、变量说明之后必须加空行。

例如：

```
int tick;
int hour;
--------------------------------空行隔开--------------------------------
hour = tick / 3600;
--------------------------------空行隔开--------------------------------
if(hour >= 59)
{
    //program code
}
```

（4）不允许把多个短语句写在同一行中，即一行只写一条语句。

例如：

```
int recData1 = 0;  int recData2 = 0;
```

应该写为

```
int recData1 = 0;
int recData2 = 0;
```

（5）if、for、do、while、case、switch、default 等语句自占一行，且 if、for、do、while 等语句的执行语句部分必须加括号{}。

例如：

```
if(s_iFreqVal > 60)
return;
```

应该写为

```
if(s_iFreqVal > 60)
{
return;
}
```

（6）当两个以上的关键字、变量、常量进行对等操作时，它们之间的操作符之前、之后或前后要加空格；进行非对等操作时，如果是关系密切的立即操作符（如->），其后不加空格。

例如：

```
int a, b, c;
if(a >= b && c > d)
a = b + c;
a *= 2;
a = b ^ 2;
*p = 'a';
flag = !isEmpty;
p = &mem;
p->id = pid;
```

A.2　注释

注释是源码程序中非常重要的一部分，通常情况下规定有效的注释量不得少于 20%。其原则是有助于对程序的阅读理解，所以注释语言必须准确、简明扼要。注释不宜太多也不宜太少，内容要一目了然，意思表达准确，避免有歧义。总之，必须加注释的地方一定要加，不必要的地方不加。

（1）边写代码边注释，修改代码的同时修改相应的注释，以保证注释与代码的一致性。不再有用的注释要删除。

（2）注释的内容要清楚明了、含义准确，防止注释二义性。

（3）避免在注释中使用缩写，特别是非常用的缩写。

（4）注释应考虑程序易读及外观排版的因素，使用的语言若是中、英文兼有的，建议多使用中文，除非能用非常流利、准确的英文表达。

A.3　命名规范

标识符的命名要清晰明了，有明确含义，同时使用完整的单词或容易理解的缩写，避免使人产生误解。

较短的单词可通过去掉“元音”形成缩写，较长的单词可取单词的头几个字母形成缩写；一些单词有大家公认的缩写。

例如：message 可缩写为 msg；flag 可缩写为 flg；increment 可缩写为 inc。

1. 三种常用命名方式介绍

（1）骆驼命名法（camelCase）

骆驼命令法，正如它的名称所表示的，是指混合使用大小写字母来构成变量和函数名字的方法。例如：printEmployeePayChecks()。

（2）帕斯卡命名法（PascalCase）

与骆驼命名法类似，只不过骆驼命名法是首个单词的首字母小写，后面单词首字母都大写；而帕斯卡命名法是所有单词首字母都大写，例如：public void DisplayInfo()。

（3）匈牙利命名法（Hungarian）

匈牙利命名法通过在变量名前面加上相应的小写字母的符号标识作为前缀，标识出变量的作用域、类型等。这些符号可以多个同时使用，顺序是先 m_（成员变量），再简单数据类型，再其他。例如：m_iFreq，表示整型的成员变量。匈牙利命名法的关键是，标识符的名字以一个或多个小写字母开头作为前缀；前缀之后是首字母大写的一个单词或多个单词组合，该单词要指明变量的用途。

2．函数命名（文件命名与函数命名相同）

函数名应能体现该函数完成的功能，可采用动词+名词的形式。关键部分应采用完整的单词，辅助部分若为常见的，可采用缩写，缩写应符合英文的规范。每个单词的首字母要大写。

例如：

```
AnalyzeSignal();
SendDataToPC();
ReadBuffer();
```

3．变量

（1）头文件为防止重编译，须使用类似_SET_CLOCK_H_的格式，其余地方应避免使用以下画线开始和结尾的定义。

例如：

```
#ifndef _SET_CLOCK_H_
#define _SET_CLOCK_H_
...
#end if
```

（2）常量使用宏的形式，且宏中的所有字母均为大写。

例如：

```
#define      MAX_VALUE     100
```

（3）枚举命名时，枚举类型名应按照EnumAbcXyz的格式，且枚举常量均为大写，不同单词之间用下画线隔开。

例如：

```
typedef enum
{
  TIME_VAL_HOUR = 0,
  TIME_VAL_MIN,
  TIME_VAL_SEC,
  TIME_VAL_MAX
}EnumTimeVal;
```

（4）结构体命名时，结构体类型名应按照StructAbcXyz的格式，且结构体的成员变量应采用骆驼命名法。

例如：

```
typedef struct
{
  short hour;
  short min;
  short sec;
}StructTimeVal;
```

（5）在本文档中，静态变量有两类，函数外定义的静态变量称为文件内部静态变量，函数内定义的静态变量称为函数内部静态变量。注意，文件内部静态变量均定义在“内部变量”区。这两种静态变量命名格式一致，即 s_+变量类型（小写）+变量名（首字母大写）。变量类型包括 i（整型）、f（浮点型）、arr（数组类型）、struct（结构体类型）、b（布尔型）、p（指针类型）。

例如：

```
s_iHour，s_arrADCConvertedValue[10]，s_pHeartRate
```

（6）函数内部的非静态变量即为局部变量，其有效区域仅限于函数范围内，局部变量命名采用骆驼命名法，即首字母小写。

例如：

```
timerStatus，tickVal，restTime
```

（7）为了最大限度地降低模块之间的耦合，本文档不建议使用全局变量，如不得已必须使用，则按照 g_+变量类型（小写）+变量名（首字母大写）进行命名。

A.4　C 文件模板

每个 C 文件模块都由模块描述区、包含头文件区、宏定义区、枚举结构体定义区、内部变量区、内部函数声明区、内部函数实现区及 API 函数实现区组成。下面是各个模块的示例。

1. 模块描述区

```
/*********************************************************************************************
* 模块名称：SendDataToHost.c
* 摘　　要：发送数据到主机
* 当前版本：1.0.0
* 作　　者：XXX
* 完成日期：20XX 年 XX 月 XX 日
* 内　　容：
* 注　　意：
**********************************************************************************************
* 取代版本：
* 作　　者：
* 完成日期：
* 修改内容：
* 修改文件：
*********************************************************************************************/
```

2. 包含头文件区

```
/*********************************************************************************************
*                                         包含头文件
*********************************************************************************************/
#include"SampleSignal.h"
#include"AnalyzeSignal.h"
#include"ProcessSignal.h"
```

3. 宏定义区

```
/*********************************************************************************************
*                                         宏定义
```

```
*****************************************************************************************/
#define  ALPHA  2048      //宏定义必须全部大写，格式为ABC_XYZ
```

4. 枚举结构体定义区

```
/*****************************************************************************************
*                                      枚举结构体定义
*****************************************************************************************/
//定义枚举
//枚举类型为EnumTimeVal，枚举类型的命名格式为EnumXxYy
typedef enum
{
  TIME_VAL_HOUR = 0,
  TIME_VAL_MIN,
  TIME_VAL_SEC,
  TIME_VAL_MAX
}EnumTimeVal;

//定义一个时间值结构体，包括3个成员变量，分别为hour、min和sec
//结构体类型为StructTimeVal，结构体类型的命名格式为StructXxYy
typedef struct
{
  short hour;
  short min;
  short sec;
}StructTimeVal;
```

5. 内部变量区

```
/*****************************************************************************************
*                                      内部变量
*****************************************************************************************/
static i16 s_iSignalSample = 0;    //信号采样值
```

6. 内部函数声明区

```
/*****************************************************************************************
*                                      内部函数声明
*****************************************************************************************/
static void SampleSignalPerSec(void *pBuf);    //每隔2ms采样一次信号
```

7. 内部函数实现区

```
/*****************************************************************************************
*                                      内部函数实现
*****************************************************************************************/
/*****************************************************************************************
* 函数名称：SampleSignal
* 函数功能：采样信号
* 输入参数：void
* 输出参数：void
* 返 回 值：void
* 创建日期：20XX年XX月XX日
* 注    意：
*****************************************************************************************/
```

```
static void SampleSignal(void)
{
}
```

8. API 函数实现区

```
/*********************************************************************************************
*                                        API 函数实现
*********************************************************************************************/
/*********************************************************************************************
* 函数名称：Task
* 函数功能：任务
* 输入参数：void
* 输出参数：void
* 返 回 值：void
* 创建日期：20XX 年 XX 月 XX 日
* 注    意：
*********************************************************************************************/
void Task(void)
{
}
```

A.5　H 文件模板

每个 H 文件模块都由模块描述区、包含头文件区、宏定义区、枚举结构体定义区及 API 函数声明区组成。下面是各个模块的示例。

1. 模块描述区

```
/*********************************************************************************************
* 模块名称：SendDataToHost.h
* 摘    要：发送数据到主机
* 当前版本：1.0
* 作    者：
* 完成日期：
* 内    容：
* 注    意：
**********************************************************************************************
* 取代版本：
* 作    者：
* 完成日期：
* 修改内容：
* 修改文件：
*********************************************************************************************/
#ifndef _SEND_DATA_TO_PC  //注意，此行代码是必需的，防止重编译
#define _SEND_DATA_TO_PC  //注意，此行代码是必需的
```

2. 包含头文件区

```
/*********************************************************************************************
*                                        包含头文件
*********************************************************************************************/
#include "DataType.h"
#include "Version.h"
```

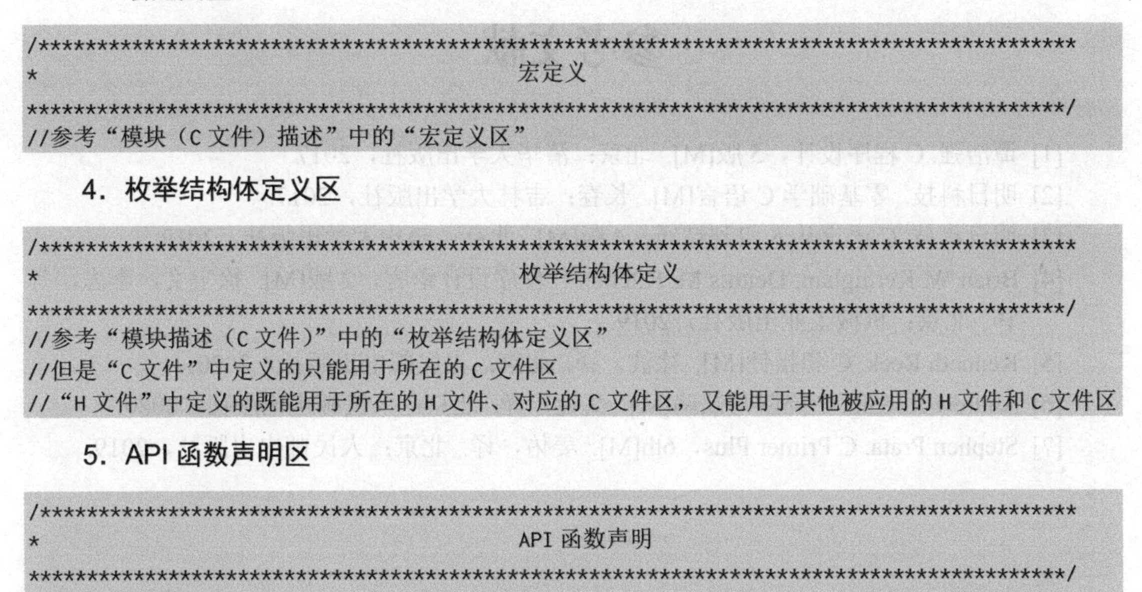

3. 宏定义区

```
/*****************************************************************************************
*                                          宏定义
*****************************************************************************************/
//参考“模块（C文件）描述”中的“宏定义区”
```

4. 枚举结构体定义区

```
/*****************************************************************************************
*                                          枚举结构体定义
*****************************************************************************************/
//参考“模块描述（C文件）”中的“枚举结构体定义区”
//但是“C文件”中定义的只能用于所在的C文件区
//“H文件”中定义的既能用于所在的H文件、对应的C文件区，又能用于其他被应用的H文件和C文件区
```

5. API函数声明区

```
/*****************************************************************************************
*                                          API函数声明
*****************************************************************************************/
void InitSignal(void);
#endif        //注意，此行代码是必需的，与#ifndef对应
```

参考文献

[1] 谭浩强. C 程序设计，5 版[M]. 北京：清华大学出版社，2017.
[2] 明日科技. 零基础学 C 语言[M]. 长春：吉林大学出版社，2017.
[3] 明日科技. C 语言从入门到精通，4 版[M]. 北京：清华大学出版社，2019.
[4] Brian W. Kernighan, Dennis M. Ritchie. C 程序设计语言，2 版[M]. 徐宝文，李志，等译. 北京：机械工业出版社，2019.
[5] Kenneth Reek. C 和指针[M]. 徐波，译. 北京：人民邮电出版社，2020.
[6] Andrew Koenig. C 陷阱与缺陷[M]. 高巍，译. 北京：人民邮电出版社，2020.
[7] Stephen Prata. C Primer Plus，6th[M]. 姜佑，译. 北京：人民邮电出版社，2019.